MA
SRI LANKA

Gehan De Silva Wijeyeratne

POCKET PHOTO GUIDE

H E L M
LONDON · OXFORD · NEW YORK · NEW DELHI · SYDNEY

HELM
Bloomsbury Publishing Plc
50 Bedford Square, London, WC1B 3DP, UK

BLOOMSBURY, HELM and the Helm logo are trademarks of
Bloomsbury Publishing Plc

First published in 2008 in Great Britain as A Photographic
Guide to Mammals of Sri Lanka
by New Holland
This edition published 2016

A catalogue record for this book is available from the British Library

Library of Congress Cataloguing in Publication data has been applied for

ISBN: PB: 978-1-4729-7598-0; ePDF: 978-1-4729-3288-4; ePub: 978-1-4729-3289-1

2 4 6 8 10 9 7 5 3

Designed and typeset in the UK by Susan McIntyre
Printed and bound in India by Replika Press

To find out more about our authors and books visit www.bloomsbury.com
and sign up for our newsletters

CONTENTS

TOPOGRAPHY AND CLIMATE ZONES

The island of Sri Lanka has lowlands along the coast that give rise within a short distance to the central hills. These rise to an altitude of more than 2,400 m above sea level. Close examination of the topography reveals that the island can be divided into three peneplains or steps, first described by the Canadian scientist Adams in 1972. The lowest peneplain lies from sea level to 30 m, the second rises to 480 m and the third reaches 1,800 m.

Sri Lanka can be divided into three regions: hill zone, low country wet zone and dry zone. The regions result from the interactions of rainfall and topography. Rainfall is affected by monsoons that bring rain during two seasons: the south-west monsoon (May–August) and the north-east monsoon (October–January). Precipitation is influenced heavily by the central hills. The monsoons deposit rain across the country and contribute to the demarcation of climate regimes.

The humid, lowland wet zone in the south-west of the island does not show marked seasons, because it is fed by both the south-west and north-east monsoons. The low country wet zone receives 200–500 cm of rain from the south-west monsoon and afternoon showers from the north-east monsoon. Humidity is high, rarely dropping below 97 per cent. Temperatures range between 27°C and 31°C over the year.

The mountainous interior lies within the wet zone. Rainfall is generally well distributed, except in Uva Province, which gets very little rain from June to September.

Temperatures in the mountains are cooler than in the lowlands and vary from chilly in the mornings to warm by noon. In the mid-elevations – such as the area around Kandy – the temperature varies between 17°C and 31°C during the year. The temperature variations during the course of a 24-hour cycle are far less varied in the mid-elevations than in the mountains. The mountains are cooler, within a band of 14°C and 32°C during the year. There may be frost in the higher hills in December and January, when temperatures fall below 0°C at night.

The rest of the country – three-quarters of Sri Lanka's land area – consists of the dry zone of the northern, southern and eastern plains. This region receives 60–190 cm of rain each year, supplied mainly by the north-east monsoon. The dry zone comprises the arid zones of the north-west and south-east, which receive less than 60 cm of rain because they are not in the direct path of the monsoonal rains.

CLASSIFICATION OF LIVING THINGS

Living things are grouped into kingdoms: animals form Kingdom Animalia and plants Kingdom Plantae. Each of these kingdoms comprises a number of phyla (singular phylum).

The plant kingdom includes the phyla Bryophyta (mosses and liverworts), Pteridophyta (ferns, horsetails, club mosses) and Spermatophyta (seed-bearing plants). Bacteria, algae and fungi may be included with the plants or in separate kingdoms. The animal kingdom in general includes the phyla Protozoa (single-celled animals), Porifera (sponges), Coelenterata (hydroids, jellyfish, corals, sea anemones), Platyhelminthes (flatworms, flukes, tapeworms), Nematoda (roundworms), Mollusca (Molluscs), Annelida (segmented worms),

Arthropoda (shrimps, spiders, insects), Echinodermata (starfish, sea urchins) and Chordata.

Phylum Chordata includes over 50,000 species with a number of shared characteristics, of which the most salient is the notochord, a strengthening chord that runs on the dorsal surface of the animal. Chordates also have a closed vascular system. This phylum has three subphyla: Cephalachordata, Tunicata and Veretebrata. The first two lack a backbone.

Vertebrates are divided into seven classes: Agnatha (jawless fish), Chondrichtyes (cartilaginous fish), Osteichthyes (bony fish), Amphibia (amphibians), Reptilia (reptiles), Aves (birds) and Mammalia (mammals).

The mammalian class is broken down into orders and, in turn, families. Orders of mammals represented in Sri Lanka include: Insectivora (shrews), Chiroptera (bats) and Carnivora (carnivores). Orders are comprised of related families — for example, Carnivora includes: Canidae (dogs), Felidae (cats), Herpestidae (mongooses), Mustelidae (otters, weasels), Ursidae (bears) and Viverridae (civets, palm-civets).

Within a family, animals are grouped into a genus and given a specific epithet to form a binomial (double name): the Golden Jackal is *Canis aureus*; the Black-backed Jackal is *Canis mesomelas*. Scientific names may change from time to time as taxonomists revise their opinions on an animal's relationship to others. Even with such revisions, scientific names are often more stable and reliable than are common names. For example, the English-language name Black-backed Jackal in Sri Lanka is commonly used for a form of the widespread Golden Jackal, *Canis aureus*, because the Sri Lankan form of Golden Jackal has a pronounced black back. Taxonomists sometimes add a third name (trinomial) to the usual binomial name to indicate a subspecies or geographical race. The Golden Jackal in Sri Lanka then becomes *Canis aureus lanka*.

To summarize, the Golden Jackal is classified as follows:

Kingdom Animalia
Phylum Chordata
Subphylum Vertebrata
Class Mammalia
Order Carnivora
Family Canidae
Genus *Canis*
Species *aureus*
Subspecies *lanka*

The binomial system of two Latin names was invented by the Swede Carl Linneaus and gives each animal a unique species name. When taxonomists reassess relationships of animals to each other, they may be moved from one genus to another or have their specific epithet changed. Occasionally one species is split into two or more, or two or more species are lumped together into one. Taxonomy remains fluid and poses a problem for both scientists and beginners. Your understanding of the behaviour of mammals will grow better if you attempt to understand family traits.

WHAT IS A MAMMAL?

Mammals are one of several classes of vertebrate animals (others are birds, reptiles, amphibians and so on). Mammals have hair on the skin, feed their young with milk and almost all give birth to live young. Mammals have four limbs, which in the case of marine mammals like whales and dolphins may be modified as flippers and in others may be reduced to vestiges. All mammals take oxygen directly from the air, which is why marine mammals have to surface to breathe and exhale. Mammals also have a heart and a blood circulatory system that allows them to maintain a constant body temperature. They can also cool themselves by using sweat glands or panting, which helps cool the body by evaporation. The ability of mammals to retain a stable temperature has enabled them to colonize a variety of habitats, from deserts to the ice packs of the Arctic and Antarctic.

MAMMAL WATCHING IN SRI LANKA

For an Asian destination, Sri Lanka is a surprisingly easy place in which to see mammals. The absolute number of species is much less than in mainland India, but the species density per square kilometre is extraordinarily high. In India, one would need to be in the core zone of Corbett National Park to see anything close to the number of species one would encounter in Sri Lanka on a good game drive in a national park such as Yala.

Of all the sites in Sri Lanka, Yala must rank as the best place in which to see mammals. To give an example, I took a game drive in Yala in September 2007 a few weeks before the manuscript of this book was due in. On just one drive we encountered the Golden Jackal, Grey Langur, Ruddy Mongoose, Sloth Bear, Wild Pig, Asian Elephant, Water Buffalo, Spotted Deer, Sambar and Palm Squirrel. A slow drive as night fell outside the park and back to the hotel produced the Jungle Cat, Gerbil, Small Civet and Indian Hare. A total of 15 species in a post-lunch session is a very good total in Asia. If we had another game drive or two to hand, we could have easily added the Toque Macaque, Stripe-necked Mongoose, Common Palm-civet, Rusty-spotted Cat and Giant Squirrel to take the tally to over 20 species – nearly a quarter of the island's mammal fauna.

Mammals are most easily seen in the dry-zone national parks such as Yala, Bundala, Uda Walawe, Wasgomuwa and Wilpattu. Some species, such as the Grey Mongoose, which is very common in the north-central province, become extremely rare in the south. Even for the mammals of the dry lowlands, taking in a mix of parks in the north-central province and the deep south helps to increase the species diversity encountered.

The endemic primates such as the Purple-faced Leaf Monkey and Toque Macaque can be found at certain sites. Talangama Wetland is not only good for critically endangered western subspecies, but also surprisingly rich in mammals generally. However, due to the high human population density, many of the mammals are nocturnal.

A visit to lowland rainforest such as Sinharaja or Kithulgala is essential to pick up some of the wet-zone species, as well as to see some of the subspecies of mammal such as the Giant Squirrel, Toque

Macaque and Purple-faced Leaf Monkey. The endemic Layard's Squirrel and the non-endemic Dusky Squirrel extend up to the highlands. A highland site such as Horton Plains will hold endemic small mammals, but these are only likely to be seen by researchers who are laying small mammal traps. Although no new species will be added by a visit to the highlands if you have already visited sites such as Sinharaja and Kithulgala, you will find that the highlands have interesting montane forms of mammals such as primates. Hakgala Botanical Gardens is very good for the highland races of the Toque Macaque and the 'Bear Monkey' race of the Purple-faced Leaf Monkey.

I also find cultural sites such as Sigirya and Polonnaruwa excellent for mammals, especially primates. I often take a game drive around the Sigiriya rock alongside the moat. My game drives regularly produce the Grey Mongoose, Toque Macaque, Hanuman Langur, northern race of Purple-faced Leaf Monkey, Palm Squirrel, Giant Squirrel and others. In the evenings I have encountered the Asian Elephant, Grey Slender Loris, Golden Palm-civet and Golden Jackal.

A NOTE ON WATCHING ANIMALS AT NIGHT

Night safaris for nocturnal animals can be a rewarding experience, but should always be done with your personal safety and the welfare of the animals in mind, and with due consideration for the concerns of state conservation agencies.

The Fauna and Flora Protection Ordinance contains several anti-hunting clauses. One of these relates to prohibitions on spotlighting animals. The original concern was relevant to hunters spotlighting an animal with the intention to shoot it. If they hid their weapons before the arrival of the game rangers they could still be prosecuted. Unfortunately for wildlife-watchers, game rangers have used the legislation to stop people from looking at animals (including birds such as nightjars) by using powerful torches and spotlights. One group of visitors that was travelling on a major road shining torches into a protected area was stopped and the vehicle impounded. Incidents such as this do little to create confidence in the state agencies, which are accused of harassing people who are enjoying wildlife with no intention to harm or kill, while poaching and illegal deforestation goes unchecked.

Incidents of over-zealous officials antagonizing wildlife lovers (rightly or wrongly) are thankfully few. Most field staff of the two main conservation agencies are doing their best to do a good job under difficult circumstances and will go out of their way to help. If you intend to look for nocturnal animals even on a public road near or passing through a protected area, speak to officials at the nearest office of the forest department or department of wildlife conservation and explain your intentions. They may even send someone along to assist you. If you are at a forest village or a cultural site, even if there is public access in the night for the forested trail you will be on, avoid suspicion by explaining to the locals what you are going to be doing. At a few sites, people have become accustomed to birders coming in search of nightjars, owls and mammals.

If you are using a spotlight or powerful torch, remember never to shine the light directly on the eyes of nocturnal animals since doing this can cause them eye damage. Always have the animals off-centre in the beam so that they are not dazzled. It is best never to use white light (to which animals are sensitive), but to use a red light (which does not bother animals). Even a weak light covered with a red filter will throw back enough eye-shine to detect nocturnal animals. The soft red light will not disturb or harm the animals. A soft red light is much better than a powerful white light for spotting and having extended views of nocturnal animals behaving naturally.

In the thorn scrub forests of the dry zone, nightwalkers are potentially at risk from Asian Elephants. An injured crop-raiding elephant is an irascible animal that may kill. Sudden encounters with Sloth Bears and Water Buffalos could also be dangerous. Kraits and vipers are active at night. Stout shoes and long trousers minimize the risk of a fatally venomous snakebite. Forest trails can also have trap guns set on them. Unless a forest trail is regularly used as a public footpath, it is best not to use it without enlisting the help of a local who is familiar with the trail.

At some of the cultural sites there is good-quality forest near the entrance gates. After a friendly chat with the security staff, I am often allowed to walk on outside the perimeter with a soft red light, with the comfort of being able to leave my vehicle under their care.

ENDEMIC MAMMALS

The following 18 species of mammal are all endemic to Sri Lanka: Ceylon Long-tailed Shrew, *Crocidura miya*, Sinharaja Shrew, Crocidura hikmiya, Kelaart's Long-clawed Shrew, *Feroculus feroculus*, Pearson's Long-clawed Shrew, *Solisorex pearsoni*, Ceylon Shrew, *Suncus zeylanicus*, Sri Lanka Pygmy Shrew, *Suncus fellowes-gordoni*, Red Slender Loris, *Loris tardigradus*, Toque Macaque, *Macaca sinica*, Purple-faced Leaf Monkey, *Trachypithecus vetulus*, Golden Palm-civet, *Paradoxurus zeylonensis*, White-spotted Mouse-deer, *Moschiola meminna*, Yellow-striped Mouse-deer, *Moschiola kathygre*, Layard's Striped Squirrel, *Funambulus layardi*, Sri Lanka Spiny Rat, *Mus mayori*, Fernando's Mouse, *Mus fernandoni*, Montane Rat, *Rattus montanus*, Ohiya Rat, *Srilankamys ohiensis*, and Long-tailed Tree Mouse, *Vandeleuria nolthenii*. Visitors are highly likely to see the two diurnal primates and the Layard's Squirrel. However, many of the other endemic mammals are small nocturnal mammals that are seldom seen other than by specialists.

KEY SITES FOR MAMMALS AND OTHER WILDLIFE

Talangama Wetland

Talangama Wetland is situated on the outskirts of Colombo. Motorable roads traverse the site, and this makes for easy access by wildlife enthusiasts. The complex of ponds, canals and paddy fields forms a rich and varied wetland site. The main target mammal species here is the critically endangered western race of the endemic Purple-faced Leaf Monkey.

The wetland still holds the Fishing Cat, Small Civet, Common Palm-civet, Indian Hare and Crested Porcupine, which are almost entirely nocturnal. Residents report Golden Jackals. During the day the Brown Mongoose is a possibility.

The wetland can be accessed from Lake Road (Wewa Para) via Akuregoda Road (DP Wijesinghe Mawatha) or Sri Wickramasinghapura Road, both of which are off the Pannipitiya Road, a few kilometres from the Parliament.

Wasgomuwa National Park

Wasgomuwa is located south of Polonnaruwa and north of the Knuckles Range and the Matale foothills. The habitat consists of riverine gallery forest along the Mahaveli and dry monsoon forest in the low foothills.

Mammals here include the Asian Elephant, Leopard, Sloth Bear, Golden Jackal, Spotted Deer, Sambar, mongooses and civets, as well as Slender Loris and Hanuman Langur. Along with Yala and Uda Walawe, this is one of the best year-round parks for Asian Elephants. A few kilometres from the park entrance beside Dunuvila Lake is Willys Safari and Dunuvila Cottage.

Horton Plains National Park

Both of Sri Lanka's second- and third-highest peaks, Kirigalpotta (2,395 m) and Thotupola Kanda (2,357 m), are in the national park. Three important rivers – the Mahaveli, Kelani and Walawe – originate within Horton Plains. The highlight for walkers is visiting World's End or Baker's Falls.

Numbers of Sambar, the island's largest deer, have soared in the last decade, with a corresponding increase in their main predator, the Leopard. Other mammals include the Wild Pig, Dusky Squirrel and the highland races of the Grizzled Indian Squirrel, Layard's Squirrel, Toque Macaque and Purple-faced Leaf Monkey (Bear Monkey). The nearest town is Nuwara Eliya, which hosts a range of accommodation.

Uda Walawe National Park

This is a popular national park because of its Asian Elephants. It is probably the only park on Earth where a sighting of wild Asian Elephants can be guaranteed. The park is a mixture of abandoned teak plantations, grassland, scrub jungle and gallery forest along the Walawe Ganga and Mau Ara. This is probably the best place to see wild herds of Asian Elephants, consisting of tightly knit family groups of up to four generations of related adult and subadult females and young. The Toque Macaque, Hanuman Langur, Spotted Deer, Wild Pig, Black-naped Hare, Ruddy Mongoose and possibly Jungle Cat may be seen in the evenings. The entrance to the park is on the B427 between Timbolketiya and Tanamanwila, near the 11 km post.

Yala (Ruhunu) National Park

Yala is undoubtedly Sri Lanka's most visited national park and the best in Sri Lanka for viewing a wide diversity of animals. It is a wonderful place with a spectrum of habitats from scrub jungle, lakes and brackish lagoons to riverine habitat. Yala National Park is divided into five blocks, of which Block 1 (Yala West) is open to the public. Yala may be closed between 1 September and 15 October. The flora is typical of dry monsoon forest vegetation in the southern belt. Plains are interspersed with pockets of forest containing species such as *Palu*, Satinwood, *Weera*, *Maila*, Mustard Tree, Neem and Woodapple.

The biggest draws in Yala are the Asian Elephant, Leopard and Sloth Bear. A recent study has shown that Yala has one of the highest densities of Leopards in the world. A game drive could yield the Black-naped Hare, Spotted Deer, Sambar, Hanuman Langur, Toque Macaque, Stripe-necked and Ruddy Mongooses, Wild Pig, Golden Jackal and others. Yala is about 40 km beyond Hambantota on the A2. Tissamaharama has a broad range of accommodation. Near the park is the Yala Village Hotel and Elephant Reach. At Weerawila is Villa Safari Hotel.

Sinharaja

The Sinharaja Man and Biosphere Reserve was declared a World Heritage Site in 1988. The reserve is arguably the most important biodiversity site in Sri Lanka, and is also internationally important for tropical biodiversity. Sinharaja comprises lowland and sub-montane wet evergreen forests with submontane Patana grasslands in the east.

A staggering 64 per cent of the tree species in this reserve are endemic to Sri Lanka. The lower slopes and valleys have remnant *Dipterocarpus* forest, with the middle and higher slopes characterized by trees of the genus *Mesua*.

Half of Sri Lanka's endemic mammals and butterflies live in Sinharaja. Visitors to the reserve are likely to see the Purple-faced Leaf Monkey and Giant Indian Squirrel.

Motorable access to the reserve is to Kudawa via Ratnapura or via Buluthota Pass from Yala or via Agalawatta and Kalawana from the west. Boulder Garden at Kalawana and Rainforest Edge near Veddagala are the nearest places with star-quality accommodation. Serious naturalists can stay at Martin's and Blue Magpie Lodge near the reserve.

Missa, Kalpitiya Peninsula and Trincomalee

These are Sri Lanka's whale-watching triangle. Trincomalee has been known for some time. In May 2008 and March 2010, the reasons why the seas south of Mirissa and those off Kalpitya Pensinsula were so good for marine mammals were explained combining scientific insight with field data. South of Mirissa is the best place in the world to see Blue Whales. Sri Lanka is in the top ten places for Sperm Whales. Pods of Spinner Dolphins, sometimes over a thousand-strong, are seen insure of the reef off the Kalpitya Peninsla. For marine mammals and seabirds, track longitude E 79 35. For more information, search the internet for my articles.

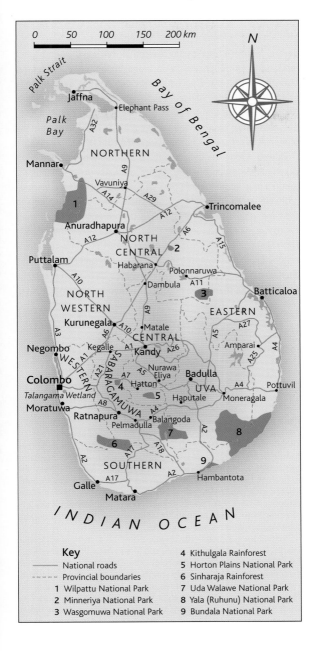

Key

National roads
Provincial boundaries
1 Wilpattu National Park
2 Minneriya National Park
3 Wasgomuwa National Park
4 Kithulgala Rainforest
5 Horton Plains National Park
6 Sinharaja Rainforest
7 Uda Walawe National Park
8 Yala (Ruhunu) National Park
9 Bundala National Park

SPECIES DESCRIPTIONS

Several mammal families are covered in this book. Due to space limitations introductions are included for only the less-known families such as shrews. The species descriptions are organized under the headings of Habitat, Distribution, Behaviour/Social Organization, Diet, Subspecies, Size and Weight and Where to See. A few lines of introductory text precede these headings. The reference to subspecies is only within the context of Sri Lanka.

Despite a long tradition of natural history study, species await description even among the 'higher' animals such as mammals, reptiles and amphibians. Some of Sri Lanka's endemic species probably vanished before science could even describe them at the time when the hill forests were cleared in the 19th century for coffee plantations. In recent years efforts led by the Wildlife Heritage Trust have seen a renaissance in Sri Lankan natural history explorations, and over 100 new vertebrate species have been described to science. Much remains to be learned about the ecology of the mammals, no matter how familiar some of them may seem.

SHREWS

These mammals look so similar to mice that most people think they belong in the same family. In fact, they are not even in the same order. Shrews are in order Insectivora in family Sciuridae (rats and mice are in family Muridae in order Rodentia). The Insectivora includes insect-eating mammals such as hedgehogs, shrews and moles. Many insectivores are not strictly insectivorous and will eat a variety of other invertebrate animals including worms, slugs, crustaceans, arthropods and even small vertebrates.

Shrews have a mouse-like appearance with a number of special characteristics. They have long, mobile snouts with sensitive hairs called vibrissae that help them locate prey. They have long tails, often exceeding their head and body length. Unlike mice, they have small eyes. Shrews have a very high metabolic rate and many have to eat their own body weight every night. Researchers capturing shrews for study thus need to check traps regularly and release shrews before they starve to death. Shrews include the smallest mammals in the world, with the Pygmy Shrew being the smallest.

HOUSE or COMMON MUSK SHREW *Suncus murinus*

This, the most common shrew, is the largest of the ten species of shrew in Sri Lanka. Depending on the subspecies, the colour varies from a bluish-grey to a blackish-brown. One pair made a nest in my map cabinet, having carefully shredded several maps in creating the equivalent of dry leaves. Like all shrews, the House Shrew has a long, pointed, mobile snout. The snout has a number of long hairs that are very sensitive to touch and enable it to detect prey in the dark. The shrew utters high-pitched squeaks, which give its presence away. Despite persecution, House Shrews can at times be surprisingly nonchalant. I was once entertaining guests when a House Shrew calmly entered the kitchen and wandered around. Knowing one of the dinner guests was likely to react hysterically, I pretended that nothing was happening.

Habitat Houses even in large cities such as Columbo.

Distribution Found throughout the island.

Behaviour/Social Organization House Shrews are nocturnal animals that may emerge at dusk. They are aggressive towards their own species and others. Most shrews eat their own body weight in the course of a night because of a high metabolic rate. Very little is known of the behavioural aspects of Sri Lankan shrews, although it is clear that olfactory communication (scent marking) plays an important role. The House Shrew is also known as the Common Musk Shrew because of the musky odour it leaves behind from scent glands on its flanks. I can often smell when one has left its calling card in my kitchen while searching for geckos and spiders.

Diet Shrews are voracious carnivores that will even attack and eat small mammals that are larger than they are. House Shrews readily

eat geckos and small lizards around a house. Much of their prey is invertebrates such as worms, insects and spiders. They will also eat mice and small snakes. Shrews are confused with mice, but being mainly carnivorous they are less troublesome as uninvited house guests and can play a useful role in keeping a house free of pests. When other food is scarce, shrews also eat vegetable matter such as fruits and grains.

Subspecies The Indian Grey House Shrew, *S. m. caerulescens*, is the largest subspecies, with a head and body length of over 14 cm. It is a bluish-grey colour. The Common Indian House Shrew, *S. m. murinus*, and Kandyan House Shrew, *S. s. kandianus*, have fur that varies from bluish-grey in the former to dark grey in the latter. They are smaller, under 12.5 cm. The subspecies intergrade, so *S. m. caerulescnes* and *murinus* and *S. s. kandianus* may be hard to separate even in the hand. The fourth subspecies (*S. m. montanus*), confined to the highlands, may be easier to distinguish as its colour is dark brownish-grey to black.

Size Head and body length 9.1–11.5 cm in *S. s. kandianus* to 14.5–15.8 cm in *S. m. caerulescens*.
Weight 25–36.5 g.

Where to See If you really want to see a House Shrew, you need to set a small mammal trap with bait. Otherwise, keep your eyes open and hope one crosses your path.

BATS

Only a few bats have been covered in this book because they are nocturnal mammals that many casual visitors to the outdoors do not get to see properly. For the benefit of the growing circle of bat watchers, an extended introduction is provided here.

Bats form a fascinating group of mammals with unique adaptations that distinguish them from other groups of mammals. They are the only mammals capable of true powered flight as opposed to gliding. Bats' forearms are modified into wings. The long, curving arms support a membrane of skin that extends from the tip of the forearm digits to the legs and usually around the legs to the space between the legs, encompassing the tail. The thighbones are twisted around so that the knees bend in the direction opposite to other mammals. A low body weight is necessary for flight, and bats have sacrificed heavy legs. They sleep upside down, which requires

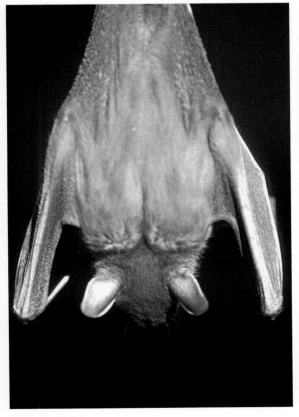

Short-nosed Fruit Bat

other evolutionary changes to both their internal organs and blood circulatory systems. Bats are famed for using echolocation, and early studies of echolocation in mammals were based on bats; they are not the only mammals to use echolocation – many cetaceans also do so.

Bats are most active during crepuscular hours (dawn and dusk). They usually sleep quietly through the day in a shaded place in a cave or tree. Bats form the order Chiroptera, which is split into two suborders: microchiroptera (around 750 species) and megachiroptera (around 150 species).

The megachiroptera are fruit bats, represented in Sri Lanka by four species: the Common Flying-fox, *Pteropus giganteus*, Dog-faced Bat, *Rousettus seminudus*, Indian Short-nosed Fruit Bat, *Cyanopterus sphinx*, and Lesser Dog-nosed Fruit Bat, *C. brachyotis*. Fruit bats use their excellent sight and sense of smell to locate food. They utter calls, but most do not use echolocation; the Dog-faced Bat is one exception. Fruit bats also lack the 'leaf' on the nose found on the echolocating bats. This leaf or tragus assists echolocation.

The flying-foxes are the best known of Sri Lanka's bats due to their large size and easy visibility. A Common Flying-fox in flight, with its wingspan of 1.2 m, is an impressive sight at close range. The animals roost in large numbers at roosts in public open spaces and at times beside main roads. Bat roosts can have complex social interactions. At times, only members of the same sex may roost together. The Dog-faced Bat is a fruit bat that roosts communally in dark caves. Like its larger cousin, the Common Flying-fox, the Dog-faced Bat is a noisy animal that can be heard some way away from the roost. The Dog-faced Bat is unusual in that it uses both sight and echolocation. In the open it relies on sight, but in dark caves and forests it uses echolocation to avoid collisions.

Suborder microchiroptera is divided into 17 families or sub-families. The exact division used between a family (scientific name ending with 'idea') and subfamily (scientific name ending in 'inae') varies from one bat taxonomist to another. Using a conventional taxonomic arrangement, Sri Lanka has eight families or sub-familes: Tomb Bats (Emballonuridae), False Vampire Bats (Megadermatidae), Leaf-nosed Bats (Hipposiderosidae), Horseshoe Bats (Rhinolophidae), Tube-nosed Bats (Murininae), Painted Bats (Kerivoulinae), Long-fingered Bats (Miniopterinae) and Free-tailed Bats (Molossidae). The infamous true Vampire Bats (Phyllostomidae or Desdomontidae) are restricted to South and Central America.

Most of the microchiropterans feed on insects taken on the wing or picked off the ground. False Vampire Bats take larger vertebrate prey such as geckos and skinks. Different bats have their own distinctive flight patterns and characteristic heights at which they fly. A good way to identify bats is by using a bat detector, which converts their high-frequency calls from outside the range of human hearing into sounds that people can hear.

Of the 89 species of terrestrial mammal recorded in Sri Lanka, no less than 31 species or over a third of all Sri Lanka's mammal species are bats. Very little is still known about many of them, and it is highly probable that new species of this nocturnal and elusive group remain undiscovered.

SHORT-NOSED FRUIT BAT *Cyanopterus sphinx*

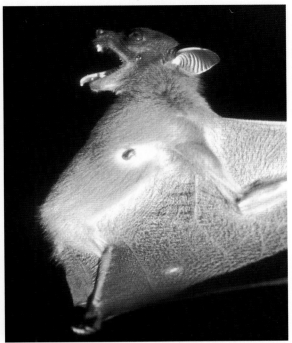

The Short-nosed Fruit Bat is one of four species of fruit bat on the island. This species and the Fulvous Fruit Bat, *Rousettus lechenaulti*, have a body length of around 10 cm (less than half the length of a Common Flying-fox). The fur on the back and nape in this species has a reddish tinge that varies in intensity with age and sex.

Habitat A frequent visitor to home gardens even in the largest of cities. It needs palm trees on which to roost. In the dry lowlands, especially in the north of the country, it roosts hanging from the leaves of the Talipot Palm, *Corypha umbraculifra*. In the wet zone it may create a roosting hollow among the seed clusters of the Kithul Tree, *Caryota urens*.

Distribution Widespread in the lowlands. The distribution needs to be re-examined, as there has been some confusion with the taxonomy. In older literature the non-endemic Lesser Dog-nosed Fruit Bat, *Cyanopterus brachyotis*, has at times been treated as a darker subspecies of Short-nosed Fruit Bat from the highlands.

Behaviour/Social Organization Roosts communally in small groups. A few individuals roost together on the same palm frond. Like

the Common Flying-fox, it orientates entirely by vision. It may also have a keen sense of smell.

Diet This species is very fond of the fruits of the Mango Tree, *Mangifera indica*, banana *Musa* spp, Soursop, *Anona muricata*, and the introduced Guava, *Psidium guajava*. Banana, Soursop and Guava are all introduced plants found in home gardens. These bats also feed on flowers.

Subspecies The Indian Short-nosed Fruit Bat, *C. s. sphinx*, found at up to 760 m, is lighter and brighter than what was previously considered a highland race, the Ceylon Short-nosed Fruit Bat, *C. s. ceylonensis*. However, the latter subspecies is now recognized as the Lesser Dog-nosed Fruit Bat (*C. brachyotis*), which is found elsewhere in Asia.

Size Head and body length 9.5–10.3 cm.
Weight Unknown.

Where to See Search for the Short-nosed Fruit Bat in groves of palm trees.

Unmistakable with its 1.2 m wingspan, this is by far the largest bat in Sri Lanka. Its face is very dog- or fox-like. The fur on its face has a yellowish tinge contrasting with the dark wings.

Habitat I have never come across a colony of Common Flying-foxes in a wooded area remote from human habitation. This may be because foraging in gardens and plantations provides a much higher nutritional yield and foraging efficiency than foraging in native forest. Common Flying-foxes are not hunted by people in Sri Lanka and may feel their colonies are more secure near people.

Distribution In lowlands up to the mid-hills, to around Kandy. A few colonies are higher up, but absent from the very highest elevations. During fruiting, they may visit higher elevations seasonally.

Behaviour/Social Organization Common Flying-foxes roost communally in a single tall tree or a cluster of tall trees. There seems to be much bickering and quarrelling going on in a colony. From time to time a few bats fly around the roost during the day, but they never feed until after dark, when they peel off one by one. If you watch from a rooftop in Colombo as the roosting fruit bats

from Viharamahadevi Park disperse, it appears that the air space over Colombo has been taken over by a swarm of fruit bats. Although they give the appearance of a flock, they forage individually, occasionally uttering a contact call in flight. They search for food using only sight. After they have landed on a tree they can move quite fast on a branch using a hand-over-hand motion.

Diet Feeds exclusively on fruits and considered a pest of fruit plantations; a frequent visitor to home gardens to feed on jak, mango and pawpaw.

Subspecies None.

Size Head and body length 26.5 cm.
Weight 0.8–1.4 kg.

Where to See Roosts on tall trees, often beside busy roads; colonies tend to be sited near human habitation.

PRIMATES

Primates are the most intelligent of mammals. They live on every continent except Australia and Antarctica, but tend to be concentrated in the tropics. Primates are of enormous interest to researchers, who study their behaviour not only for its intrinsic interest but also because it may shed light on the evolution of social behaviour in humans.

The order of Primates comprises 15 families, including the Hominidae (incorporating modern humans in the genus *Homo*). Primates are split into two suborders: Prosimii (lemurs, lorises, galagos and tarsiers) and Anthropoidea (monkeys, apes and humans). Anthropoidea is divided into Platyrrhini (New World monkeys) and Catarrhini (Old World monkeys, apes and humans).

Apes and humans form the family Hominoidea and Old World monkeys are in Cercopithecoidea. Old World monkeys are further divided into two subfamilies: Cercopithecinae (baboons, macaques, guenons and relatives) and Colobinae (langurs and colobus monkeys).

The suborder Prosmii and the subfamilies Cercopithecinae and Colobinae (in the suborder Anthropoidea) are represented in Sri Lanka. The total number of five species in Sri Lanka represents a fair cross-section of the primate evolutionary tree. The island has a high

Southern Purple-faced Leaf Monkey

density of primate species per 1,000 sq km. At certain sites – such as Polonnaruwa Archaeological reserve – the density of individuals is also high. The number at Polonnaruwa is artificially high due to a combination of protection and the extra food discarded by visitors. The ease with which primates can be found and observed makes Sri Lanka a very good location for primate studies. In fact, a study by the primatologist Dr Wolfgang Dittus on the Toque Macaques is one of the longest-running field studies in the world and spans four decades.

Prosimians are considered the most ancient order of primates. In Sri Lanka it was believed until recently that only one species, the Grey Slender Loris, *Loris lydekkerianus*, was present. However, the loris in the wet-zone lowlands is distinctive, both in terms of behaviour and in size and appearance, and has recently been named as a new species, the Red Slender Loris, *Loris tardigradus*. The Red Slender Loris is smaller and more active than its relative. It can move rapidly, almost scampering on branches. The highland race in Horton Plains National Park and the isolated population in the Knuckles Range are also potential candidates for a split.

Hanuman Langurs occasionally congregate into large troops

Red Slender Loris

Red Slender Loris

Lorises have a tapetum in the eye behind the retina. This is a reflective layer of cells that reflects light back through the retina, effectively giving the eyes a second pass of the light to improve night vision. Cats have a similar tapetum, but have evolved it independently. The best way to locate lorises is by their eye shine at night. Anyone participating in loris watching should use a weak red light (lorises are not sensitive to red light). LCD lights should never be used because they may be damaging to animals (see also page 7).

Lorises (like all wet-nosed primates, the Strepsirrhines) rely heavily on olfactory clues (smell) in marking territories and signalling readiness to breed. They have large nostrils and scent glands that facilitate their use of scent. One of the digits in the forearms has a claw called the toilet claw. This finger is used in various grooming activities, such as scratching or cleaning the auditory canal. Lorises also have a modified tooth with comb-like serrations. The toothcomb is used for grooming the fur. Lorises have a range of vocalizations, but for the most part a shrill whistle is all that is heard. This is easily overlooked as an insect's call. This and lorises' unobtrusive behaviour may explain why people are often not aware of the presence of these cryptic animals in a forest. The montane race went undetected for over seven decades until primatologist Dr Anna Nekaris found it again after a three-day search in the Horton Plains.

The two subfamilies of Old World monkeys live in Asia and Africa. The endemic Purple-faced Leaf Monkey, *Presbytis senex*, and the Hanuman Langur, *Presbytis entellus*, are in the Colobinae. The endemic Toque Monkey or Toque Macaque is in the Cercopithecinae. All of the Old World monkeys have 32 teeth, as do humans. Other primates have more teeth. The reduction in teeth reflects a change in diet and facial structure that has taken place in the Old World monkeys. The size of the incisors relative to the canines indicates their diet. Cercopithecines have larger incisors, which they use for biting into fruit. Their molars are more flattened for crushing and grinding hard seeds. The colobines are more adapted to a folivorous diet and their incisors and molars do not show the same adaptations. They

Above: Toque Monkey

*Right: Toque Monkey
(montane)*

*Below: Southern Purple-
faced Leaf Monkey*

Top and above: Toque Monkeys

have a sacculated stomach where digestive-aiding gut bacteria reside. The bacteria break down the hard cellulose walls of plant cells and can even detoxify poisonous seeds and other plant matter.

The cercopithecines have salivary compounds in their cheek pouches that break down toxins in plant material. The cheek pouches also help to deal with feeding competition. One can see Toque Macaques that have literally stuffed their faces. The colobines lack cheek pouches. The cercopithecines also differ from the colobines in having arms and legs of a similar length; the colobines have much longer legs and usually have a long tail. The colobines are typically shy and unlikely to show aggression towards people, but at least one

Top and above: Red Slender Loris

habituated population of Hanuman Langurs has shown aggression. The usually gentle Hanuman Langur can be a fearsome sight when it is gnashing its teeth and showing its canines. Toque Macaques should always be treated with caution, especially where people feed them, which leaves them with the allusion that humans are subordinate.

Primates have a number of adaptations, including stereoscopic vision, opposable thumbs and a collarbone that gives increased mobility. Humans possess all of these characteristics and we may overlook the evolutionary advantages conferred over other mammals due to these adaptations. It is difficult when studying wild primates not be struck by how similar they are to us.

Hanuman Langur chasing juvenile jackal

Threat yawn of a Hanuman Langur

Primates are among the best-studied of Sri Lankan mammals, thanks to Drs Anna Nekaris, Wolfgang Dittus, Jinie Dela and their students (from Sri Lanka and overseas), working in collaboration with Sri Lankan scientists and students. Studies of the Sri Lankan diurnal primates show that they have male-dominated hierarchies with an alpha male, but studies on Toque Macaques show that the troops are formed on matrilineal lines. The females provide the stable core for troops. Toques are hierarchical and the young of high-placed females inherit a high status in the troop. With so much opportunity to observe primates at close quarters, and a wealth of fieldwork, Sri Lanka is a great place for watching primates.

RED SLENDER LORIS *Loris tardigradus*

The Red Slender Loris is smaller than the Grey Slender Loris and has relatively small ears and a longer muzzle. It was named for its reddish colour. The dark markings around the eyes are more circular (or semi-circular) in shape compared with the ovular or teardrop-shaped markings on the Grey Slender Loris.

Habitat Prefers primary or secondary rainforest. The home range of lorises varies from 1.5 to 15 hectares. They are unlikely to survive in very small forest fragments as they do not travel great distances and are vulnerable to cats and dogs where forests are fragmented due to housing developments.

Distribution Confined to forests in the wet zone. Populations of the Red Slender Loris can survive in extensive areas of home gardens mixed with wooded patches. Until 2003, there were reliable records of the Red Slender Loris from sites such as Talangama on the outskirts of Colombo.

Behaviour/Social Organization A recent radio-tracking study conducted by Lilia Bernede (Oxford Brookes University and University of Ruhuna) has begun to shed light on an animal about which relatively

little was known by gathering data on a small population that was followed for one year. The study also investigated the vocal repertoire of this species. It appears that it has more variants of the common whistle call than expected. Slender Lorises stalk prey and grab their food with lightning speed once within strike range. On-going studies will shed light on home range sizes for males and females, but as an approximate guideline a loris may need up to 10 hectares. Red Slender Lorises seem to be more reluctant than Grey Slender Lorises to travel along the ground, making it important for them that vines or overlapping branches connect trees. Loss of habitat has eliminated Red Slender Lorises from near Colombo, although I have friends who still have lorises visiting their forested gardens beside Bolgoda Lake.

Diet Lorises mostly eat insects, but will also consume other small animals. In captivity they sometimes eat fruits.

Subspecies *L. t. tardigradus* in the lowland wet zone. *L. t. nycticeboides* has a denser and woollier coat, shorter limbs and ears that are covered in fur. This subspecies is confined to the cloud forests in and around Horton Plains National Park. Further studies may show it to be a distinct species.

Size Head and body length 19.2 cm.
Weight 140 g.

Where to See Wet-zone forests such as Sinharaja and Morapitiya.

GREY SLENDER LORIS *Loris lydekkerianus*

There are two subspecies of the Grey Slender Loris. The descriptions below are for the Northern Grey Slender Loris, *L. l. nordicus*, which lives in the dry lowlands. The Grey Slender Loris has a greyish coat and yellow pigmentation on the ears, eyelids and muzzle. The dark shapes around the eye are oval or teardrop shaped. The species is bigger than the Red Slender Loris.

Habitat Lightly wooded scrub forest with closed canopy in the dry lowland. Often seen at forest edges, where it presumably benefits from the edge effect and access to a wider range of prey. This species had adapted to forest disturbance and fragmentation better than its relative, as it is more likely to come down and scramble along the ground at times to cross main roads or pass from one forested patch to another. It will also make its way along barbed wire fences to traverse from one forest patch to another.

Distribution Distributed widely on the dry lowlands, but seems to be more abundant in the north central province, especially around sites such as Sigiriya and Polonnaruwa. Mannar Island seems to have a particularly high density of lorises. Curiously, there are no firm sight records of it in Yala, but I have spoken to safari jeep drivers who have seen it in Bundala and Debera Wewa.

Behaviour/Social Organization Lorises are entirely nocturnal and easily overlooked unless you go out looking for them and are familiar with their call. The most frequent of their calls is a thin, drawn-out whistle that probably lasts at the most a few seconds and that can be mistaken for the call of an insect. Other calls described are a chatter and a scream. The Grey Slender Loris may have a different social structure from that of the Red Slender Loris, although this will need to be confirmed by field studies. In Polonnaruwa, these lorises are observed to sleep communally, with as many as eleven individuals gathering on occasions.

Diet Grey Slender Lorises are primarily insectivorous, but have also been observed to eat other animals such as snails and lizards.

Subspecies *L. l. nordicus* is found throughout the dry lowlands and has yellow pigmentation on the ears, eyelids and muzzle. The subspecies is endemic to the island. *L. l. grandis* has a woolier coat and shows little yellow pigmentation. It is slightly heavier. This subspecies lives in rainforests in the central mountains extending from Kandy to the Knuckles Wilderness. The subspecies is endemic to the island. It is possible that the form found in the Knuckles may turn out to be a distinct species.

Size Head and body length 22 cm.
Weight 220–280 g.

Where to See The scrub forests around Sigiriya and Polonnaruwa are good. The Teak Forest, a small hotel in Sigiriya, has lorises visiting its grounds. Lorises are also seen within a short drive of the Vil Uyana Hotel.

TOQUE MACAQUE or TOQUE MONKEY *Macaca sinica*

Above: M. s. aurifrons Top: M. s. sinica Right: M. s. opisthomelas

The Toque Macaque lends the impression that it is very common because it occurs in significant numbers in places such as the Cultural Triangle sites, where visitors gather – but it is not as common as might be thought based on this evidence.

Habitat Toques like forest cover near water bodies. They occur near human habitation, but need trees for roosting.

Distribution The three subspecies occupy forested habitats all over the island.

Behaviour/Social Organization Toque Macaques have complex social organizations. Dominant alpha males monopolize mating with receptive females. The females form stable matrilines, which may be a more influential factor in the long run in Toque Macaque society than that of alpha males. The females belong to castes with different degrees of social status.

Diet Omnivorous – a wide range of plant and animal matter.

Subspecies The colour and length of the hairs radiating from the cap (toque) is used to distinguish the subspecies. This is not always easy to see in the field between the dry-zone race and the race in the wet lowlands. The race *M. s. sinica* lives in the dry lowlands. The tips of the hairs radiating from the cap are a pale buff colour. The wet-zone race *M. s. aurifrons* is generally darker and has reddish or yellowish tips to the hairs on the toque. The montane race *M. s. opisthomelas* has long hairs radiating from the cap, with the tips straw coloured. The upperparts and sides lack the reddish-brown of the lowland races.

Size Head and body length 41–52 cm; tail 52–56 cm.
Weight 5–11 kg.

Where to See The dry-zone race at Anuradhapura, Polonnaruwa and Sigiriya; wet-zone race at Udawattakale in Kandy; montane race at Hakgala Botanical Gardens.

HANUMAN or GREY LANGUR *Semnopithecus priam*

The Hanuman Langur is a graceful, dark-faced monkey with a sinuous long tail. It may be mistaken for the Purple-faced Leaf Monkey, which also has a dark face. In Sri Lanka, the Hanuman Langur is easily told apart from the Leaf Monkey by the pointed bonnet on its head. The Leaf Monkey also has more extensive white whiskers on its chin and cheeks.

Habitat Scrub forests and riverine forests in the dry lowlands.

Distribution Confined to the dry lowlands.

Behaviour/Social Organization Hanuman Langurs can at times form large troops exceeding 50 individuals, but most of the time they live in troops of fewer than 20 individuals, with a dominant male. They are usually very shy in most places apart from a few temples where they have got used to people. Unlike Toque Macaques, habituated Hanuman Langurs are generally not aggressive towards people. Hanuman Langurs are often found in the company of Spotted Deer: the deer benefit from the leaves and fruits dropped by the langurs; the langurs benefit from the extra pairs of eyes.

Diet Herbivorous. A narrow diet confined to tender leaves and fruits.

Subspecies None.

Size Head and body length 54.5 cm; tail 75 cm.
Weight 9–16 kg.

Where to See The archaeological sites of Anuradhapura, Polonnaruwa and Sigiriya.

PURPLE-FACED LEAF MONKEY *Trachypithecus vetulus*

T. v. monticola

The Purple-faced Leaf Monkey is a shy monkey that never seems to be quite at ease with people. The montane race is habituated to people at Hakgala Botanical Gardens, but will never approach people with the boldness of Toque Macaques. Not crashing away at the approach of people seems to be as far as they get to being habituated. On the outskirts of Colombo, in the Talangama Wetland and around Bolgoda Lake, they are almost daily visitors to home gardens. People have mixed reactions to the monkeys. Unfortunately, many see them as a pest that raids their fruits and so attempt to chase them away. There are even reports of barbaric people having shot these legally protected monkeys for no good reason. Amazingly, at the time of writing a small troop holds out in the heart of central Colombo, centred around Buller's Road, but as it is separated from other populations it is unlikely to survive in the long term.

Habitat The natural habitat for the Leaf Monkey is tall forests. In urban areas it is still found in reasonable numbers because it feeds on jak and mango, which are popular trees in home gardens. Leaf Monkeys are particularly fond of ripe jak fruit, and most good sightings are when they have gathered on a jak fruit. For some years now the law has prohibited the felling of jak trees for timber. The price of jak timber had climbed very high, and there was a risk that almost all of the good stock of jak trees in people's gardens would be felled for timber. The purpose of this law was to help poor families, for whom jak trees are a source of food. What was intended as an anti-poverty measure has turned out to be a great boon for the conservation of primates and other native wildlife.

Distribution Distributed widely throughout the island where tall forests are found. It is absent from the northern peninsula.

Behaviour/Social Organization Generally found in small family troops of up to half a dozen or a dozen individuals. One troop in Talangama has up to 40 individuals, but this is unusual; most troops in Talangama have around half a dozen individuals. The standard social structure is an alpha male hierarchy with several females and young in the troop. Loose associations are established between young males, who form bachelor troops. They may collaborate to topple an alpha male.

Although these monkeys are usually quite shy, I once had an alpha male bound towards me from the trees and stop a few feet away gnashing his teeth and barking furiously. My offence was to turn into my private reserve and park under a troop in which a female had a baby less than a week old. I am now careful to note where the monkeys are before I park.

Diet Largely folivorous (leaf eating), but also eats fruits. The habit of raiding mango, jak and coconuts creates conflict with some people.

Above: T. v. nestor

Below: T. v. vetulus

Subspecies The Southern Purple-faced Leaf Monkey, *T. v. vetulus*, has a well-defined, large, silvery white rump patch that extends down the thighs. At certain angles when the monkey is seen in a dimly lit forest it may look as though it is wearing a pair of white shorts. The tail has more white than that of the other subspecies. The southern race is found south of the Nilwala River. This is the subspecies seen at Sinhraja and in the rainforests around Galle. In the Western Purple-faced Leaf Monkey, *T. v. nestor*, the rump patch is large but diffused. In adult males the rump patch can look similar to that of some individuals of race *T. v. vetulus*, but the tail never becomes so white. This subspecies is classified as Critically Endangered. In the Bear Monkey, *T. v. monticola*, the long fur imparts a shaggy appearance. The tail is short and stout. The Bear Monkey is confined to the highlands, where it is seen around Hakgala and Horton Plains National Park. The Northern Purple-faced Leaf Monkey, *T. v. philbricki*, is somewhat paler overall than the other forms, with short hair and a long, slender tail. Some vocalizations are different from those of the western race, which lives futher south. The northern form is likely to be seen in the cultural sites around Anuradhapura, Polonnaruwa and Sigiriya.

Size Head and body length 49 cm (female) to 68.4 cm (male).
Weight 5.2–11.4 kg in males of some subspecies.

Where to See The western race of this species is most easily seen at the Talangama Wetland, and also around Bolgoda Lake. The montane form is easy to see at Hakgala Botanical Gardens. Look for the northern subspecies around Sigiriya. The southern race is best seen at Sinhraja, where a troop near the barrier gate is reasonably tolerant of people.

GOLDEN or BLACK-BACKED JACKAL *Canis aureus*

Next to the Leopard and crocodile, the Golden Jackal is probably the most important high-level carnivore regulating populations of prey

Above: Golden Jackal scent marking

in the lowland jungles, especially in the dry zone. The Sri Lankan subspecies *C. a. lankae* is also known as the Black-backed Jackal, and looks noticeably different from the form of Golden Jackal that lives in mainland India.

Habitat Hunts in open habitats such as grasslands, as well as in scrub forest.

Distribution Found throughout the island, but less prevalent in the higher mountains. The Golden Jackal still holds out in extensive areas of marshlands such as Muthurajawela, about 20 km north of Colombo. It is seen at the Talangama Wetland, which is even closer to Colombo, and in areas adjoining the Bolgoda Lake. Sightings by visitors are likely to be almost entirely in the national parks in the dry zone.

Behaviour/Social Organization Golden Jackals hunt in small packs, although quite often they are seen as a pair. They have an interesting social structure in which the alpha pair breeds and the

others in the pack are helpers. It is possible that helpers benefit by learning the skills of parenthood. In the national parks in the dry zone, Golden Jackals may be encountered at almost any time, although they are more likely to be lying up in a shaded thicket during the heat of the day. Where they are persecuted, they become strictly nocturnal.

Diet Omnivorous. Golden Jackals have a wide diet ranging from lizards and birds to small mammals. They are also scavengers of dead animals and will eat fruits and berries.

Subspecies The subspecies *C. a. lankae* is endemic to the island.

Size Head and body length 68.8–77.5 cm; tail 20–24.1 cm.
Weight 7–15 kg.

Where to See Yala National Park is best.

SLOTH BEAR *Melursus ursinus*

Long, powerful claws and uneven black fur, which can look patchy on many adults, characterize the Sloth Bear. It has a long, pale muzzle. This is the only species of bear found on the island. Bears are among the most dangerous of animals to encounter and many a villager has a mauled face bearing witness to a bear attack. Sloth Bears frequently attack the heads of victims, and may sometimes rear up on their hind legs to do so.

Most Sloth Bears in Sri Lanka will not stalk and kill a person with the intent of eating. Bear attacks almost always arise when a bear is startled or with young. I have spoken to many villagers and field staff of national parks and reserves. Opinion is divided as to the best form of defence during a bear attack: some advocate standing your ground and shouting and clapping; others argue that you should beat a hasty retreat.

Habitat Scrub and monsoon forests in the dry lowlands. Requires denning sites such as tree hollows or rock outcrops with cool and dark recesses.

Distribution Lives throughout the dry lowlands. The distribution of bears does not coincide with the distribution of termites, one of its favourite foods. It is a puzzle as to why it is absent from the wet zone. In the past, Sloth Bears were known to make movements up into the hills in search of food during times of extreme drought.

Behaviour/Social Organization Sloth Bears are usually seen singly. It would seem that males part company with females after mating. Females are often seen accompanied by their young (usually two, but sometimes three).

Sloth Bears may spend the day in communal sleeping sites. On one occasion I came across a mother lying on her back with her feet in the air and a cub on the soles of her feet. I could almost imagine her juggling the infant. As my jeep came around the bend, she got to her feet and one of the cubs clambered onto her back for a ride. Within a matter of seconds the family had vanished. Bears are not territorial; males and females share their home ranges with other male or female bears. Sometimes both sexes are encountered while they patrol or forage for food.

Old battle wounds enable certain individuals – such as 'One Eye' – to be recognized in places such as Yala National Park, where they have

become reasonably tolerant of safari vehicles. It is almost impossible to see a Sloth Bear outside of the national parks (such as Yala and Wilpattu), because they are nocturnal. Wasgomuwa National Park is reputed to have the highest density of Sloth Bears, but even there it is very difficult to obtain a daytime sighting. Radio collar studies in Wasgomuwa by researcher Shyamala Ratnayeke and others are beginning to shed more light on the social behaviour of Sloth Bears. The study found that male and female Sloth Bears had home ranges of 2.2 and 3.8 sq km respectively. There was also considerable overlap between the home ranges within and between sexes.

Diet Like all species of bear, the Sloth Bear varies its diet according to the availability of its food and consumes a wide range of plant and animal matter. However, it is on the evolutionary road to semi-specialization and has lost a few front teeth – this enables it to cup its licks and create a suction pump to suck in termites. Powerful claws enable it to rip down the strong termite mounds of baked mud to feed on the insects. During the dry season, when termite mounds are as hard as concrete, the bear switches to fruits and berries, grubs and other insects. As a rule, a Sloth Bear will not hunt animals actively, although it will happily steal from the kill of another carnivore. The fruit of the Palu Tree, *Manilkra hexandra*, is a favourite of the bear. During July, when the trees are in fruit, Sloth Bears can be seen gorging themselves.

Subspecies The subspecies *M. u. inornatus* is restricted to Sri Lanka.

Size Head and body length 1.36–1.58 m; tail 13–16.5 cm; height at shoulder 24 cm (female); 36 cm (male).
Weight 68.2 kg (female); 104.5 kg (male).

Where to See Yala and Wilpattu National Parks are the best options all year round. The time when the Palu is in fruit, in July, is an especially good time for daytime sightings.

OTTER *Lutra lutra*

The Otter is an enigmatic animal. Many Sri Lankans are probably not even aware that this mammal occurs in Sri Lanka. Those who are are probably less than pleased – they may know it because it cleaned out the fish in their ponds in one overnight raid. On land it seems a little awkward because of its short feet, which are webbed in adaptation for aquatic hunting. It has a thick, sausage-shaped body with a tail that is thick at the base. The Otter is uniformly chocolate brown on the upperparts and pale underneath. It is very graceful in water.

Habitat Lives wherever unpolluted streams, rivers, ponds and lakes are found throughout the island. On a visit to Mannar Island, I once saw one crossing a busy main road in the town near Medwachchiya, but it will only pass through urban areas and will not settle down unless there is a quite a wide extent of natural green area together

with clean water. It holds out in the suburbs of Colombo, which are enmeshed in the remains of a once-extensive network of wetlands. Sarinda Unamboowe has watched Otters while using the gym at the Water's Edge golf course and country club development in the Kotte Marshes.

Distribution Lives throughout the island. I have seen it in the well-known Arrenga Pool at Horton Plains National Park and photographer Namal Kamalgoda had an obliging individual who swam around while he took photographs from within his vehicle.

Behaviour/Social Organization Largely nocturnal, although I have had a few daytime sightings, which may be due to predation pressures from domestic dogs and persecution from people who may mistake it for a mongoose (the latter is not popular as it kills domestic poultry).

Diet Fish comprise the large part of its diet, but it also eats a range of aquatic animals, from invertebrates such as molluscs and freshwater crabs to amphibians. Being a sizeable carnivore it will also eat small mammals such as rodents and water birds if the opportunity presents itself.

Subspecies The subspecies *L. l. nair* is considered to be restricted to southern India and Sri Lanka.

Size Head and body length 55.4–61.4 cm; tail 35–40.1 cm.
Weight 3.9–5.3 kg.

Where to See Hunas Falls Hotel near Kandy has over the years had otters that swim on the lake during the day from time to time.

VIVERRIDS (CIVETS, PALM-CIVETS AND MONGOOSES)

Viverrids comprise the genets, civets and mongooses. With 66 species in 37 genera, this is the largest family of carnivores distributed in the Old World tropics. It is an ancient family of mammals first appearing in the fossil record in the Lower Oligocene. Fossil records show that their basic body form has not changed much for around 40–50 million years. They are generalized carnivores; some will eat a variety of plant and animal matter. A skull of a mongoose assigned to the genus *Herpestes* has been dated to 30 million years ago and shows it to be the oldest surviving carnivore genus. All of Sri Lanka's mongooses are in this genus.

Three subfamilies of Viverridae are found in Sri Lanka. The subfamily Viverrinae is represented by one species in Sri Lanka: the Small Civet. Civets are cat-like nocturnal animals that prefer to hunt on the ground, although they are good climbers. Subfamily Paradoxurinae has two species in Sri Lanka, although there will soon be three species recognized with the endemic Golden Palm-civet being split into two species. Palm-civets are arboreal and nocturnal. The Golden Palm-civet seldom descends to the ground. It consumes a lot of fruits and berries, as well as meat. Subfamily Herpestinae has four species of mongoose in Sri Lanka. They are short-legged, long-bodied and not cat-like. They are mainly diurnal and likely to be seen easily by visitors to wilderness areas. Some taxonomists place the three subfamilies into two families: Viverridae (civets and palm-civets) and Herpestidae (mongooses).

Scent plays a very important role in communication for the viverrids, with vocalizations being limited to a few bird-like yelps. All viverrids have well-developed anal glands with which they deposit scent marks in their territory.

Ruddy Mongoose

COMMON PALM-CIVET or TODDY CAT
Paradoxurus hermaphroditus

Above and below: Juvenile Common Palm-civets

This is a common mammal, but one that few people see because of its nocturnal habits. When seen in the dark it can seem jet black, but the fur is more greyish than black and seems to fit somewhat loosely on its body. It is a long-bodied and long-tailed mammal with short feet. It has a pale patch on the face that breaks up its outline, a feature common to many nocturnal mammals.

Habitat It needs wooded patches in the wild. In urban areas it has adapted to well-wooded gardens, even in the heart of big cities such as Colombo.

Distribution Found throughout the island.

Behaviour/Social Organization Usually seen singly. The young seem to be taken care of by the mother. There are 3–4 young in a litter. The young utter a bird-like twittering call. In the wild the species will sleep and breed in tree holes or crevices in rocky outcrops. In cities it seems perfectly happy to find a dry corner under a roof. My house in Colombo 8 has palm-civets in residence from time to time. They make their presence felt when they bound across roofs and jump from one

level to another, landing with a loud thud. I once watched a mother jump about two body lengths across from one roof to another with a youngster held in her mouth. Palm-civets can shuffle backwards in a straight line without any difficulty. This is probably an adaptation to enable them to reverse out of tree holes and rock crevices, which they investigate in search of food.

Diet Eats a wide variety of food, from small mammals such as rats and mice to insects and fruits. It is often seen climbing to the tops of fruiting trees to eat fruits. Its droppings, frequently found strewn on rocks and logs (and garden paving around my house), often contain the digested seeds of the Kithul Tree, *Caryota urens*. It is said to be fond of stealing the juice of the tree that is 'tapped', hence the local name of Toddy Cat.

Subspecies Sri Lanka has just one subspecies, *P. h. hermaphroditus*, which is darker than some of the Asian mainland forms.

Size Head and body length 47.5–56 cm; tail 43.2–47 cm.
Weight 2.3 kg (female); 3.3 kg (male).

Where to See Night safaris in wooded areas.

GOLDEN PALM-CIVET *Paradoxurus zeylonensis*

The Golden Palm-civet is not likely to be confused with any of the other palm-civet species because of its uniform reddish colour. It is slightly smaller than the more common Common Palm-civet, and is overlooked because of its arboreal habits. It is usually seen only during a night walk in a forested area by shining a torch beam (use a weak red light) into the crown of tall trees. I have found it to be fairly self-assured when it is on a tree as it will often watch the observer calmly and seems to be in no hurry to move away.

Habitat Found in tall forests. It does not seem to be averse to occupying tall trees close to human habitation. It is absent from areas where large stretches of forest are no longer available.

Distribution From the lowlands to the mid-hills.

Behaviour/Social Organization Often seen singly or in pairs. It seems to live all its life up in the trees, and will sleep in a tree hole.

Diet A mix of fruits, berries and small animals such as birds, lizards, amphibians, insects and so on. Its main food probably comprises native fruits, and it will visit home gardens to feed on jak, mango, bananas and other cultivated fruits.

Subspecies The single species described is endemic to Sri Lanka. However, the animal in the dry zone is probably a different species from that found in the wet zone: the split into two species is in the process of being published.

Size Head and body length 46.5–55.9 cm; tail 42.2–48.9 cm.
Weight 1.7–2.3 kg (female); 3.3 kg (male).

Where to See Tall forests around the Sigiriya moat and Martin's Simple Lodge in Sinharaja (read the section on watching animals at night, see page 7).

SMALL or RING-TAILED CIVET *Viverricula indica*

The Small Civet is an unmistakable animal with a black-and-white body and a tail with black-and-white rings. It is one of the most common nocturnal animals to be seen when driving at night along forested roads out in the country.

Small Civet pup

Habitat An animal of forests, but it has become habituated and will visit houses and explore dustbins for food. Quite a few bungalows inside national parks have a regular civet that will come in and turn over the bin to look for scraps of food, although it is much shyer of people than the Common Palm-civet (which will take up residence in the roofs of houses).

Distribution Found throughout the island from the lowlands to the highlands.

Behaviour/Social Organization The Small Civet is usually seen singly. Like all civets and palm-civets, it marks its territory with droppings on exposed rocks and tree trunks. It often hunts on the ground, although it is equally comfortable climbing trees in pursuit of prey.

Diet Prefers to hunt for small mammals such as rodents, hares and mouse-deer. It will also eat fruits, berries and roots, and is more carnivorous than the three Sri Lankan palm-civet species.

Subspecies Not described at subspecies level in Sri Lanka; also lives elsewhere in eastern and southern Asia.

Size Head and body length 52.4–58.9 cm; tail 33.8–37.8 cm.
Weight 2.1 kg (female); 2.8 kg (male).

Where to See Often seen on night drives on public roads running through forested areas. The time around midnight, when traffic is at its lowest, will offer the best chances of seeing it.

INDIAN GREY MONGOOSE *Herpestes edwardsii*

The uniform grizzled grey coat without any contrasting markings easily identifies the Grey Mongoose. It is smaller than the other mongooses, and the eyes are reddish.

Habitat Occurs in grasslands and grassy verges, as well as lightly wooded country.

Distribution Found throughout the island, but most common northwards from the north central province. In the cultural triangle area of the dry lowlands, it is impossible to drive around for a day without seeing one. In contrast, it is very scarce in the south. I have had just a handful of sightings out of several hundred game drives in Yala National Park.

Behaviour/Social Organization It tends to be a solitary animal, although at times pairs are seen. It is extremely wary, and usually dashes across a road. At certain places, like the moat of the Sigiriya, it has become accustomed to people and may sit on a wall sunning itself. But if you stop and point a camera at it, it will more often than not feel ill at ease with the attention and move away. It is very diurnal in its habits.

Diet It is omnivorous and will take fruits and roots. However, it prefers animals from juicy invertebrates to small rodents.

Subspecies The subspecies *H. e. lanka* described from Sri Lanka is unique to the island.

Size Head and body length 35–45.6 cm; tail 25.5–45.6 cm.
Weight 0.8–1.4 kg.

Where to See The moat around Sigiriya early in the morning is a good place to find one on the prowl.

INDIAN BROWN MONGOOSE *Herpestes fuscus*

The only species the Indian Brown Mongoose could be confused with is the Ruddy Mongoose. It can be separated from it by the lack of a black tip to the tail, and the end of the tail not being carried in an upturned manner. A close view reveals that the coat lacks the grizzled appearance of the Ruddy. Unlike the Ruddy, which has become fairly tolerant towards safari vehicles in the national parks, the Brown and the Grey Mongooses remain extremely wary and will take flight if you stop to pay them any attention. I have had a Brown Mongoose playing in my garden during the day, but it would bolt out of sight if I ventured out with a camera. It is one of the hardest mammals to photograph, even though it lives in even the most densely inhabited cities.

Habitat A very adaptable animal that can live in the heart of Colombo as well as in densely wooded forests.

Distribution Found throughout the island up to the highlands.

Behaviour/Social Organization The Indian Brown Mongoose is usually seen as a solitary animal. It can be active during the day, but seems more so at night. In places where it is not persecuted it will visit rubbish dumps during daylight. Mature adults are rarely seen during the day in cities such as Colombo, but in a less urban setting – such as the Talangama Wetland – wary adults may occasionally forage during the day.

Diet Omnivorous, the mainstay of its diet being animals ranging from worms and grubs to small mammals such as rodents. It also takes fruits and berries occasionally.

Subspecies Four subspecies have been described based mainly on variations in colour. The Ceylon Highland Brown Mongoose, *H. f. flavidens*, lives in the wet zone from the western province in the lowlands to the highlands. It is considered a dark and stocky

subspecies. The Western Ceylon Brown Mongoose, *H. f. rubidor*, extends from the western province to the south and is described as having shorter fur. The Northern Ceylon Brown Mongoose, *H. f. maccarthiae*, is found in the northern parts. The published variation in its colour ('...dorsal contours only indistinctly speckled with blackish...', Pocock 1941) seems to be based on just one individual. The Mannar Ceylon Brown Mongoose, *H. f. siccatus*, lives in the Mannar–Aripu area in the arid north-western coastal strip. Its colour variation ('general colour nearly uniform sandy or strawlike...', Pocock 1941) is once again based on a single individual.

There is no doubt that populations, for example on the Mannar–Aripu stretch, may show a size and colour different from those of a population in the Horton Plains. However, given the degree of individual variation within a given population of Brown Mongooses, a number of different populations could be established, and the subspecific variations in the Brown Mongoose probably need a critical contemporary evaluation.

Size Head and body length 32.9–44.3 cm; tail 23.6–31 cm.
Weight 0.6 kg (female); 1.2 kg (male).
The wide variation in measurements is because races *H. f. flavidens* (smaller) and *H. f. rubidor* (larger) are included together in the measurements.

Where to See Visits by Brown Mongooses to the compost heap at the Blue Magpie Lodge can be watched at a safe distance from a balcony. Hunas Falls Hotel has one or more animals that patrol the grounds. Other than at the Blue Magpie Lodge, sightings are fleeting and seldom permit an opportunity for photography.

RUDDY MONGOOSE *Herpestes smithii*

The Ruddy Mongoose is a common mongoose of the lowlands in the southern half of the island. It may be confused with the Brown Mongoose – which also has a uniformly brown colour on the body and tail – but the Ruddy has a characteristic black tip to the tail. The Ruddy also carries its tail rudely curled upwards as if engaging in an act of 'giving you the finger'.

Habitat Forested areas.

Distribution The Ruddy Mongoose is found throughout the island, but is displaced by the Grey Mongoose in the north and the Brown Mongoose in the west.

Behaviour/Social Organization Often in pairs. Somewhat shy: tends to melt into the forest when encountered on a game drive, although sometimes a pair will walk nonchalantly past a parked safari vehicle. Seen during the day in national parks such as Yala.

Diet Carnivorous. It also scavenges the kills of other animals, and is a common visitor to kills made by Leopards.

Subspecies The subspecies *H. s. zeylanicus* is unique to Sri Lanka.

Size Head and body length 39.4–45.3 cm; tail 33–36 cm.
Weight 1–2 kg (male bigger and heavier than female).

Where to See Yala National Park.

STRIPE-NECKED MONGOOSE *Herpestes vitticollis*

This species, the largest mongoose in Asia, has a distinctive black stripe on its neck, starting from behind the ear and extending towards the top of the forelegs. The black tail tip is thick and fluffy. Mature adults have grey on the head and show a rusty brown diffuse band around the rump. The legs are black. No other mongoose in Sri Lanka is as colourful.

Habitat It is seldom seen in areas away from forest. It is most common in dry-zone scrub forest, but also lives in rainforest.

Distribution Stripe-necked Mongooses live throughout the island up to Horton Plains National Park. They are scarce in the north of the island.

Behaviour/Social Organization Equally likely to be seen in a pair and as a solitary individual. Frequency of sightings of these animals can be variable: on some visits to Yala National Park I do not have even one sighting, yet on other visits I seem to see them every other game drive. I have often watched a pair digging in soft, damp soil or sandy soil to ferret out grubs. They can be so absorbed in digging that they take no notice of safari vehicles that gather around to observe them. I have on occasion watched them for over half an hour digging beside a jeep track. Outside national parks, they are extremely shy and almost impossible to see. They are largely diurnal.

Diet Omnivorous. They will eat fruits, berries and roots. Because of their large size they can also hunt small mammals such as mouse-deer and hares. Birds, birds' eggs, and invertebrates and their larvae are all on the menu.

Subspecies The species is restricted to southern India and Sri Lanka. The nominate subspecies *H. v. viticollis* is found in Sri Lanka and southern India; a second subspecies is confined to North Kanara, India.

Size Head and body length 46.5–49.9 cm; tail 30.5 cm.
Weight 1.7 kg (female); 3.1 kg (male).

Where to See Yala National Park is a reliable site. They are seen less frequently at Horton Plains National Park.

CATS

Fishing Cat

The 36 species of cat in the family Felidae live around the world except on the continents of Australia and Antarctica. Of all carnivores, cats are the most specialized as meat eaters. Their short gut does not allow them to switch to vegetarian sources of food. Meat is always in relatively short supply, and cats have a number of biological adaptations for efficient hunting. Cats evolved to their present state around 40 million years ago.

Broadly speaking, cats employ two techniques for hunting: patrolling in search of prey and lying in wait. The background colour of a cat's coat reflects the general colour of its habitat. In a forested environment, where the trees create shadows and dappled light, the coat may be striped or spotted. The underneath of the tip of the tail may be white – as in the Cheetah and Leopard – as a visual cue for cubs following their mother.

In some cats, polymorphism occurs (more than one type of colouration may be present). The black panther is an example of such a cat. A black panther is a melanistic Leopard, whose coat is rendered almost black because of a recessive gene. In Malaysia, the dark form of the Leopard is more common than the yellow and spotted form. It may be that the dark colour confers a competitive advantage for Leopards in rainforests.

Leopard subadults engaged in mock fighting

Jungle Cat – melanistic form (left) and normal form (right)

Cats have more developed senses than humans for sight, smell and hearing. For most species of cat, vision is the most important sense. However, what is meant by better vision needs some clarification. Humans actually have a higher visual acuity – the ability to resolve detail under good light – than cats do. Cats, on the other hand, have more sensitive vision, especially under low levels of lighting. A cat's retina has, like that of a human, both rods and cones. Cones detect colour. Rods are not sensitive to colour, but are more sensitive to low light. The high density of rods in a cat's eyes allows it to see in light where a human would not be able to see at all. The eyes of cats are about six times more sensitive to light than human eyes. Rods dominate most of a cat's retina, with a smaller cone-rich area in the centre. It is likely that colour does not play a large part in a cat's life.

Cats and other nocturnal animals have another mechanism for night vision. Behind the retina is a membrane called the tapetum lucidum. This reflects light back through the retina, increasing the sensitivity during low light. This is what causes the eyes of nocturnal animals to glow brightly when an artificial light is shone on them. To see well in both dull and bright light, cats control the diameter of the pupil using muscles in the eye. This allows them to increase the pupil's diameter and increase the amount of light captured at night.

As many cats are adept climbers, being able to judge distances accurately is important. Cats have the most developed binocular vision of all carnivores (but not as good as humans and other primates).

Smell plays a small part in hunting, but it is in communicating with other members of the same species that it becomes a very important tool. Cats will scent mark their territories using a combination of methods. The most common method is spraying urine. Another is using their faeces as a marker. Yet another is 'scratching', where scent from inter-digital glands is mixed in.

Visitors to Sri Lankan national parks may encounter scent-marking Leopards at any time of day, although it is most likely to be at

Rusty-spotted Cat

dawn or dusk, when Leopards are closing or beginning their session of activity.

Many mammals perform an action known as flehmen: the lips are flared back exposing the teeth, and air is inhaled sharply. The incoming air is tested by a urine analysis in the Jacobson's organ in the brain. A duct on the upper jaw connects to nerve centres on the brain and enables sampling to take place. This organ is absent in humans. Male Leopards patrolling their territories perform flehem on urine markings done by females. This helps the male identify the female's breeding condition. The males also test urine to learn details of intruders into their territory.

Most cats lead a solitary existence. Lions are the only truly social cats, although bachelor groups of Cheetahs may hunt together. The mating strategy of cats such as Leopards can be understood in terms of an individual selfishly attempting to maximize the passage of his genes. As only the fittest of males can control and dominate a territory, it follows that only the fittest genes are passed on from one generation to another.

JUNGLE CAT *Felis chaus*

An elegant cat, the Jungle Cat reaches the size of a small dog. It has a uniformly sandy brown coat with a greyish tone and pointed ears. The underparts are lighter. The tips of the ears have black hairs, giving them a blackish rim. The first impression on a daytime sighting is often of a jackal-like animal.

Habitat Dry-zone scrub forest is the haunt of this species. The regular sightings in Uda Walawe National Park demonstrate that it may have a preference for hunting in tall grasslands.

Distribution It seems to be restricted to the dry lowlands.

Behaviour/Social Organization Most sightings are of a single animal. The species may have the typical felid social structure of a male controlling a home range that encompasses the home range of several females. Very little is known about the Jungle Cat's behaviour in the wild. It can be seen hunting early in the morning and early in the evening, although it is probably nocturnal where there is a human presence. Despite its name, the Jungle Cat is more of a grassland cat. This is one way of niche partitioning with other wild cats that will maintain overlapping territories. The different prey sizes taken by the various cats also help reduce competition, although a large cat such as a Leopard will also opportunistically take the same prey as a Jungle Cat.

Diet Predominantly small mammals such as hares, mouse-deer and rodents, and birds and their eggs.

Subspecies The subspecies found in Sri Lanka, *F. c. keelarti*, also lives in southern India.

Size Head and body length 61–64.8 cm; tail 22.9–28.2 cm; height at shoulder 45.72 cm.
Weight 5–8.2 kg.

Where to See Uda Walawe National Park on an evening game drive is the most reliable site. I have also come across it on public roads through dry-zone forests around midnight, when vehicular traffic is virtually absent.

RUSTY-SPOTTED CAT *Prionailurus rubiginosus*

The smallest wild cat in Sri Lanka is also just about the smallest wild cat in the world. It is close in size to a small domestic cat, and may easily be overlooked as a feral cat. The Rusty-spotted Cat can be identified by the white stripes on the head, rusty spots on a grey-brown body and unmarked tail. The facial pattern is similar to that of the Fishing Cat.

Habitat It needs forest cover. Occurs in tall forests in the wet zone as well as in thorn scrub forests in the dry zone. In both types of forest it is often encountered along forest trails. I have come across it hunting besides motorable dirt roads running though dry-zone forests.

Distribution Lives throughout the island from the lowlands to the highlands.

Behaviour/Social Organization Very little is known of this animal in the wild in Sri Lanka. Most sightings are of solitary individuals. It probably has a social system like that of most cats, with males only pairing up for mating. It does not seem to be very shy. The Vil Uyana Hotel in Sigiriya had a Rusty-spotted Cat that began visiting the hotel's premises soon after it was opened. I have pulled up near one in the night and it has not been too alarmed at the presence of a vehicle. It probably lives quite close to people in villages adjoining forest, but is never found where good forests are not present. It climbs trees to hunt or rest, but most sightings are of cats patrolling on the ground.

Diet Mostly small mammals such as mice and shrews, and birds. It is known to raid hen houses.

Subspecies The subspecies *P. r. phillipsi* is confined to Sri Lanka.

Size Head and body length 38.3–47 cm; tail 20.8–25 cm.
Weight From 1.1 kg (female); 1.6 kg (male).

Where to See A night safari on a forested road may yield a sighting, but stay on public roads and read the tips on watching animals at night (see page 7).

FISHING CAT *Prionailurus viverrinus*

This is a large wild cat – a full-grown male is the size of a dog. In other parts of its range it has been known to kill children. An adult can be quite self-assured. Sunela Jayawardene, Sri Lanka's leading environmental architect, reports seeing a Fishing Cat standing its ground and snarling at her and her companions at her bungalow in Wasgomuwa when they were between the cat and its intended meal.

The animal has two white stripes extending up from the eye to its forehead. It also has two prominent white patches on the cheeks. The area around the muzzle is white on both the upper and lower jaws

(a feature shared by the Rusty-spotted Cat). The fur is greyish, with the sides having blotches loosely forming longitudinal stripes. On the dorsal area, the longitudinal stripes are more distinct. Its tail is thick and appears heavier at the rear end.

Habitat Aquatic habitats that are bordered by suitable forest cover. Most lakes, ponds, canals and streams with wooded areas harbour a population of these animals. Even heavily built-up areas such as those on the outskirts of Colombo still have Fishing Cats because sufficient habitat remains of the once-extensive wetlands.

Distribution From the lowlands to the highlands, although absent from northern parts of the island.

Behaviour/Social Organization Very little is known about the behaviour of this cat in the wild. On-going research may shed some clues on its social organization. It is mainly nocturnal where it is disturbed by the presence of people, but otherwise will hunt by day. Daytime reports are regular even in suburban areas, where the Fishing Cat occurs in good numbers. Residents in Talangama regularly report daytime sightings. Dunvila Cottage in Wasgomuwa was visited by an individual with a habit of attempting daytime raids on poultry.

Diet It will feed on small animals and fish. Because of its size, most medium-sized animals are potential prey.

Subspecies Not described at the subspecies level from Sri Lanka; widely distributed in Asia.

Size Head and body length 65.8–85.7 cm; tail 25.4–28 cm; height at shoulder over 35.56 cm.
Weight 6.4 kg (female); 11.8 kg (male).

Where to See In aquatic habitats that are criss-crossed by roads (for easy access to observers). Talangama Wetland and Kotte Marshes yield sightings throughout the year.

LEOPARD *Panthera pardus kotiya*

Sri Lanka is the best place in Asia – and indeed one of the best in the world – to see the Leopard. It has become a symbol of Sri Lanka for its potential as a destination for wildlife travel. A number of factors contribute to the good viewing of Leopards in Sri Lanka. The first is that it is the top predator. Unlike in Africa, it does not have to compete with lions and hyenas. In India it has to compete with tigers. Both lions and tigers will kill Leopards.

Being Sri Lanka's top terrestrial predator imbues it with a certain confidence. As a result, cubs and subadults are very relaxed during the day and will often stay out within sight of a convoy of admiring visitors in safari vehicles.

Leopard subadult, relaxed in absence of higher predators

The second factor for good viewing of Leopards is the density of their population. Research by the late Ravi Samarasinha, a Leopard researcher, showed that the average density of Leopards in his study area of Block 1 in Yala (Ruhunu) National Park was as high as one per square kilometre. The average density is a higher figure than that based on home ranges alone because in a given area, in addition to the territory holders, there are also cubs, subadults and transients that do not hold territories.

The high average density of Leopards reflects the high prey density. In sites such as Yala this is a result of past human activities, which created a mosaic of grasslands and scrub with numerous waterholes. It is almost as if a deliberate attempt has been made to increase the population of Spotted Deer, the main prey of the Leopard. The protection from hunting (at least to a large degree) is a third factor. Fourthly, viewing is easy because of the terrain of the park: fairly short grassland and scrub with plenty of rocky outcrops for Leopards to climb.

Habitat Leopards are very adaptable creatures. Worldwide they have adapted to environments from harsh deserts to alpine habitats just below the snow line. In Sri Lanka they are best seen in the scrub jungles in the dry lowlands, but are also found in the rainforests of the wet lowlands and in the cloud forests of the highlands.

Distribution Leopards were once found throughout the island. Due to intense human activity they have now been extirpated from the northern peninsula and the populous western province. Where there is suitable forest cover, they continue to hold out. Leopards come to within 100 m of Nuwara Eliya town because of the cloud forests cloaking the Mount Pedro Forest range. Even in recent years there have been sightings of Leopards coming down to drink water from Kandy Lake. Such a sighting is extremely unusual, but it does

demonstrate that Leopards still hold out in the forested ridges around the town of Kandy.

Behaviour/Social Organization
Like most cats, Leopards employ a single male/multi-female social system. The size of a Leopard's territory depends on the availability of prey. In Sub-Saharan Africa, a male Leopard's home range may encompass 100 sq km. In Yala, a male's home range is 16–20 sq km. This encompasses the home ranges of 3–4 females whose home ranges may be 2–4 sq km. The core of its home range, the territory, which it will defend rigorously, is smaller than the home range. Leopards scent mark their territory. This enables them to avoid actual physical combat, which may be fatal to one or both combatants. Early in the morning or late in the evening, adult Leopards can be encountered walking along dirt tracks and scent marking. They often spray their urine upwards so that it coats the undersides of leaves where it is likely to persist longer. They will

also use faeces and scratch marks to announce their presence. A rival Leopard is able to assess the sex and condition of another Leopard from its scent marks.

Male Leopards guard their territories and thereby maintain exclusive access to the females within them. When a female is at the peak of receptivity to mating, the male will stay with her for several days and mate with her. It is the female that initiates courtship by flirting with the male. The male takes no part in bringing up the young; after mating he parts company with the female. The young are termed cubs until the age of about six months and termed subadults until about 18 months. At this stage they are totally dependent on the mother.

At about 18 months of age, male cubs especially begin exploratory forays to seek a territory. At this time the males are the size of the mother, and mother and cub are often mistaken for a pair of adults. Most daytime sightings of Leopards are of subadults. After about two years the males disperse and become nocturnal in habit. This is a very dangerous time for a young adult as existing territory holders will fight them to the death. I have come across at least one body of a young adult that had challenged a territory holder with fatal consequences.

Males tend to push out further away from their mother's territory, while young females carve out a territory near the mother. The instinct of Leopards to push out in search of a territory has resulted in Bundala National Park once again being occupied by Leopards after being bereft for several decades.

Diet Leopards are carnivorous and eat a range of mammals, from shrews and rodents to buffalo. In national parks such as Yala, the Spotted Deer is the preferred prey. In Horton Plains National Park, the Spotted Deer is absent but Sambar populations have grown in recent decades and Leopards have switched to them as their preferred prey. Where large prey is unavailable, a Leopard's diet switches to smaller prey such as Muntjac, mouse-deer, primates and Land Monitors; almost any animal can be preyed on. Where the habitat available to Leopards is shrinking due to the presence of people, Leopard's have also begun to take domestic dogs, a tactic that creates conflict with the local community.

Subspecies The subspecies *P. p. kotiya* found in Sri Lanka is endemic to the island. The present consensus is that worldwide the number of subspecies of Leopard may be seven or less (see box).

Size Head and body length 1.05–1.42 m; tail 77.5–96.5 cm.
Weight 29.1 kg (female); 77.3 kg (male).

Where to See Yala (Ruhunu) National Park is by far the best site in Asia for seeing Leopards. There is a 90 per cent chance of seeing Leopards if you take five game drives. This is a cautious estimate because you can have a run of bad luck and fail to see a Leopard after five game drives. But there have been many occasions when I have had time for only one game drive with a visiting journalist and the Leopards have obliged.

How Many Leopard Subspecies?

Dr Sriyanie Miththapala undertook a DNA study of Leopards for her doctoral dissertation. She and her research colleagues reduced the number of described subspecies of Leopard from twenty-seven to just seven. The work by her and her colleagues found four morphologically distinct groups: African, Asian, Sri Lankan and Javan. Twelve subspecies described from Africa were reduced to one subspecies, *P. p. pardus*. Six subspecies in Central Asia were reduced to one subspecies, *P. p. saxicolour*. Three subspecies on the Indian subcontinent were reduced to one subspecies, *P. p. fusca*. The subspecies of Leopard in Java, *P. p. melas*, and that in Sri Lanka, *P. p. kotiya*, were found to be distinct in both molecular and morphological analyses. This accounts for five subspecies. Due to a lack of individuals, two other subspecies – the Amur Leopard, *P. p. orientalis*, and North Chinese Leopard, *P. p. japonensis* – are retained to making the total of seven subspecies presently accepted.

Two mature bull elephants fighting for dominance

ASIAN ELEPHANT *Elephas maximus*

The largest terrestrial mammal in Asia is taken for granted locally because it is so familiar. But it is unusual in so many ways. Which other mammal uses its nose to pick up food?

The Asian Elephant has other adaptations that are not obvious to humans: for example, it uses infrasound to communicate. Infrasound consists of long waves of low-frequency sound that travel distances of several kilometres. Individuals and clans keep in touch with and track and follow the movements of each other inaudibly to humans. I once watched a male in Yala listening intently to another male, using his feet to feel the sound! Other safari vehicles drove past, bored with a bull standing rigid, while I waited for what culminated in a serious battle between two bulls.

The soles of the feet of both the Asian Elephant and the African Elephant are lined with onion-like layers of nerves called Pacinian

corpuscles, with a gelatinous layer. It is thought that these nerves enable an elephant to listen to seismic waves. By stamping its feet an elephant may generate seismic waves for other elephants to listen to tens of kilometres away.

One characteristic of the Asian Elephant that distinguishes it from the African Elephant is its concave rather than convex back. It is also smaller. Another, easy distinguishing feature is the large, triangular ears of the African Elephant compared with the considerably smaller, rounded ears of the Asian Elephant. In addition, the tip of the Asian Elephant's trunk has one protuberance or finger while that of the African has two.

Habitat The Asian Elephant is a forest elephant – unlike most populations of African Elephant, which are adapted to savanna. The

African Elephant uses its ears as an air-conditioning unit. The cooling effect of passing blood through its ears can result in a temperature drop of 19°C. The Asian Elephant, with its smaller ears, has a much lower tolerance to heat and is unlikely to emerge on open grasslands before the cool of the evening. It is probably because of the heat that it is mostly nocturnal. Pressure from human activity would also reinforce nocturnal behaviour. Nevertheless, groups of elephants are easily seen engaged in daytime foraging, especially when it is cool, in Sri Lanka's national parks.

African Elephants of both sexes grow tusks, but in Asian Elephants only males – and only a small proportion of them – carry tusks, which are elongated incisors. In Sri Lanka it is estimated that only 7–8 per cent of the males carry tusks. The term tushes is used to describe the small tusks that some males have. The same term is used for the small, tusk-like milk teeth that many elephants are born with but shed by the time they are two years old. The only other teeth that elephants have are molars.

Distribution Only a couple of hundred years ago, Asian Elephants ranged from the scrub forests and grasslands of the dry lowlands to the cloud forests of Horton Plains. Now, only a few pocketed groups remain in the mid-hills around the Peak Wilderness. Significant numbers of Asian Elephants remain only in the dry lowlands.

Behaviour/Social Organization The mother and calf are the basic unit of elephant society. A bond group of Asian Elephants usually consists of related animals with subadult male siblings, female siblings, mother, aunts and cousins. An adult female will suckle a calf that is not its own. These females are termed allo mothers. When a male Asian Elephant reaches puberty, it is pushed out of the bond group. The young males may form bachelor herds. As they grow older, they become solitary bulls. Bond groups combine with other bond groups to form clans. A sexually mature male may at times join clans, seeking to mate with females in oestrus. Clans may occasionally coalesce with other clans to form herds. In Sri Lanka herds are often seen when

Asian Elephants gather during times of drought around a large water body. Herds are not stable social structures, and eventually they disperse into the smaller clans.

Bull elephants periodically undergo a biochemical change called musth. During this time they are in peak breeding condition and are pumped up with hormonal secretions in their circulatory system. They are markedly more aggressive: a larger bull not in musth will give way to a smaller bull in musth. Bulls in musth release a secretion from the temporal glands near the ears and often deliberately spray their pungent urine on their hind legs. They make no secret of being in musth. Captive bulls in musth are chained to prevent them from being a danger to people.

Adult elephants are very protective of their young and will flank the youngsters when crossing a road. Very young babies often remain under the mother's belly. A newly born calf is able to walk almost immediately. I have observed babies that were so young that their skin was pink. The adult females guarded them so carefully that they

were often hidden from view, surrounded by a wall of protective adults.

Radio-collaring studies have shown that adult bulls may range over as much as 140 sq km. Unfortunately, Asian Elephants are becoming increasingly penned in behind electric fences. The human–elephant conflict has escalated tremendously in recent years, and each year a few score people and elephants are killed. Asian Elephants face a grim future in Sri Lanka unless policy makers strive to keep open vast areas for wildlife to roam.

Diet Asian Elephants need a combination of grassland in which they graze, and leaves from shrubs and trees to browse. Bacteria in their stomachs break down the cellulose in the plant matter. Baby elephants are born without this bacteria. I have watched baby elephants ingesting the dung of adults. This must be an instinctive response that enables them to obtain the microflora they need in their stomachs for digestion. Asian Elephants consume around 150 kg of vegetable matter in a day.

Subspecies The subspecies *E. m. maximus* found in Sri Lanka is the largest of the Asian Elephant subspecies. Deraniyagala described a subspecies, *E. m. vilaliya*, known as the Marsh Elephant, from the riverine habitats of the large Mahaveli River in the north-eastern plains. However, this subspecies has not gained wide acceptance. Modern DNA studies by Dr Pruthuviraj Fernando have shown that the genetic distance between different populations of Asian Elephant in Sri Lanka is greater than that between the Sri Lankan elephants and those on the Indian mainland. Elephants clearly do not make migratory movements as such between the north and the south of the island. This does not mean that the case for wildlife corridors is weakened. Whether it is for elephants or small animals such as butterflies, jungle corridors are needed so that through generations of random dispersals, gene flow is maintained between different populations, so reducing inbreeding.

Size Height at shoulder for males around 1.1–1.36 m; females around 0.98–1.01 m. Phillips records a large male shot in Tamankaduwa (of the unaccepted race *E. m. vilaliya*), which was 1.2 m at the shoulder. **Weight** 3,000 kg.

Where to See Visitors to Uda Walawa National Park are guaranteed to see Asian Elephants. Sri Lanka is the only place where visitors can be certain of seeing wild Asian Elephants on just a single game drive. Wasgomuwa and Yala (Ruhunu) National Parks are also good places. The Gathering (see pages 82–3) at Minneriya and Kaudulla National Parks is not to be missed.

The Gathering

Small islands are not supposed to have large animals. Someone forgot to tell this to the Asian Elephant, the largest terrestrial mammal in Asia. Not only does it live in Sri Lanka, but also the largest concentration of these Asian giants, a seasonal gathering, takes place on this island. Every year the Gathering takes place on the receding shores of the Minneriya Tank in the north-central province of Sri Lanka. As the dry season fastens its grip on the dry lowlands, leaves wither and fall in the dry monsoon forests, and waterholes evaporate into cakes of cracked and parched mud. Asian Elephants must move on in search of food and water. The Asian Elephants, sometimes numbering over an awe-inspiring 300, converge onto the receding shores of Minneriya Tank. It is the highest concentration of wild Asian Elephants on Earth.

The Gathering at Minneriya is a wonderful opportunity for both wildlife enthusiasts and casual travellers to observe the social dynamics of Asian Elephants. Clans of related elephants coalesce into herds when they converge onto Minneriya in a common quest for food, water, cover and mates. The smaller herds group into even larger herds, sometimes numbering over 100 Asian Elephants. Adult bulls mix freely, using their trunks to test the air for adult females who are receptive. Bulls tussle for dominance and calves play with each other.

The Minneriya Tank or reservoir is an ancient manmade lake constructed by King Mahasen in the 3rd century AD. Many centuries ago, these lowlands were farmed for agriculture by an ancient civilization whose mastery of hydraulics was remarkably sophisticated. Today, the ancient reservoir fills during the north-east monsoon and gradually shrinks as the dry season fastens the lowlands in a torpid grip. As the waters recede, lush grassland sprouts, attracting Asian Elephants in search of food from as far away as the jungles of Wasgomuwa and Trincomalee.

Key Facts

When should I visit?

The Gathering peaks during the months of August and September. The locals will know whether the herds are gathered at Minneriya National Park or whether the nearby Kaudulla National Park offers better viewing at a particular time. You should be guided by local advice and be flexible as to which of the parks you visit.

Why is it called the Gathering?

Because that is what it is. It is a seasonal movement of Asian Elephants and not quite a migration in the sense of what biologists mean by a migration.

How should I visit?

Choose a reputable tour operator who can make your arrangements for accommodation, park entry fees and safari jeep hire. Hotels in the neighborhood can also make arrangements for jeep safaris.

What else can I visit on the same occasion?

Minneriya – the focus of the Gathering – is at the centre of one of the richest areas for culture and archaeology in Sri Lanka. The magnificent ancient cities of Anuradhapura and Polonnaruwa, the rock fortress palace of Kasyappa at Sigiriya and the Golden Rock Temple of Dambulla are all within a half-day's excursion from Minneriya. Wildlife and culture enthusiasts may like to visit the Ritigala archaeological and forest reserve. Polonnaruwa and Sigiriya are outstanding archaeological sites that are also excellent for watching primates. Several of Sri Lanka's finest hotels are located within half an hour to an hour's drive of Minneriya.

WILD PIG *Sus scrofa*

The Wild Pig is generally a very shy animal because it is hunted illegally for its flesh. In protected areas in the dry lowlands, it is more confiding and at times may gather into herds or sounders numbering over 30 individuals. At game lodges, the animals become quite bold. They are fearless and will readily take on a Leopard that threatens a herd. I once observed a sounder of Wild Pigs stalked by a Leopard. The young and adult females quickly gathered into a creche with adult males (boars) throwing a defensive ring around them. Four males then approached the Leopard, which fled up a tree. The males stood guard for a while, allowing the females and young to move away.

Habitat Found wherever dry-zone scrub or wet-zone forest provides it with forest cover.

Distribution Found throughout the island up to the highlands.

Behaviour/Social Organization Occurs in large herds in the national parks in the dry lowlands. Small herds or solitary animals live in the wet-zone forests. The males develop formidable tusks that can inflict heavy damage on a hunter. Adult males are often seen alone, and it seems likely that they join the herds of females and young only to breed.

Diet Omnivorous. Causes a lot of damage to home gardens and plantations by uprooting plants for rhizomes. Will readily scavenge meat. I once came across a herd tearing apart and eating a Spotted Deer fawn. How the fawn died was not clear.

Subspecies None.

Size Head and body length 1.8 m.
Weight 100 kg.

Where to See National parks.

DEER

Sri Lanka has four species of deer in the family Cervidae and two species in the family Tragulidae. Of the four species in the Cervidae, it is not clear whether the Hog Deer, *Axis porcinus*, is native or whether it was introduced. It is restricted to coastal swamplands in the south-west. The males in the deer family that adopt a harem system develop impressive antlers. Older individuals show more tines (branches) on the antlers. These are grown and shed seasonally. The size of the antlers is probably an indication to females of the fitness of a male. The antlers are used in combat. Because they are a heavy load to bear, males shed them once the breeding season is over. I have seen deer nibbling shed antlers, perhaps to ingest calcium and other minerals. During growth the antlers are covered with velvet, which is dense with blood vessels, and the males are referred to as being in velvet.

Most species of deer have alpha males that control a harem of females. This way a single male monopolizes access to females. The male expends a significant amount of energy shepherding his harem and guarding it from other males. Spotted Deer and Sambar employ the harem system. Indian Muntjac or Barking Deer are usually solitarily or in pairs. Deer have facial glands. Muntjac have pronounced facial glands, and scent marking may play an important role in maintaining pair bonds and territories.

Large herds of Spotted Deer are restricted to national parks

Deer scent mark by rubbing their facial glands against plants such as grasses and bushes. They also have glands on their feet. These glands may be used to deposit scent when they make scrapes on the ground. Deer may additionally use dung middens to mark their territory.

Two species of mouse-deer are presently recognized in Sri Lanka: the White-spotted Mouse-deer, *Moschiola meminna*, and Yellow-striped Mouse-deer, *Moschiola kathygre*. I have coined these common names from the description by Groves and Meijaard in 2005, where they proposed splitting mouse-deer into two species. They also speculate that the mouse-deer in the highlands may be another new and different species.

The species in India was previously considered to be the same species as the one in Sri Lanka, but is now regarded as a distinct species: Indian Mouse-deer, *Moschiola indica*. The mouse-deer in the genus Moschiola are closely related to the chevrotains in the genus *Tragulus*, found in South-East Asia, and the African or Water Chevrotain, *Hyemoischus aquaticus*.

The mouse-deer in the family Tragulidae are considered to be an ancient group going back to the Miocene epoch. They are thought to be a sister group of the Ruminantia. Mouse-deer are not true tree-climbers, but they are able to clamber up trees that are at a slant and covered in creepers. They do so to escape from terrestrial predators such as dogs. Mouse-deer are not very vocal, but will give a bark-like call when alarmed.

WHITE-SPOTTED MOUSE-DEER *Moschiola meminna*

The Indian Mouse-deer, *M. indica*, is described by Groves and Meijaard (2005) as having the upper row of spots forming a continuous stripe on the shoulder and breaking down into spots halfway along the body. The White-spotted Mouse-deer of Sri Lanka is characterized by the upper row of spots not fusing to form a continuous stripe even on the shoulder. The colour of both the Indian Mouse-deer and White-spotted Mouse-deer is a dull brown.

Habitat Found in the scrub forests of the dry zone. Has adapted to gardens in villages.

Distribution Found in the dry zone of Sri Lanka.

Behaviour/Social Organization Usually seen singly or in pairs. Largely nocturnal, but I have seen animals active during the day inside national parks.

Diet Feeds on grasses, herbaceous plants, tender leaves and fallen fruits and berries.

Subspecies None described. Recently split as a new species.

Size Head and body length 50.5–56 cm; tail 2.25–3 cm.
Weight 3.8 kg (female); 3.1 kg (male).
These measurements are based on Phillips (1980) and the range encompasses both species of mouse-deer.

Where to See National Parks such as Yala and Wilpattu may occasionally yield a sighting as one dashes across the road.

YELLOW-STRIPED MOUSE-DEER *Moschiola kathygre*

Groves and Meijaard (2005) describe the colour of this species as being more ochre-brown and warmer than in other species. The spots and stripes are more yellowy than white. The two longitudinal rows of elongated spots on the flanks are fused into 'tolerably complete longitudinal stripes' with an elongated spot row between them and two more spot rows above them.

Habitat Forests in the wet zone. It often seeks refuge in home gardens as well.

Distribution Found in the wet zone up to the mid-hills. The range extends in the south-east to Katagamuwa in Yala, where the dry-zone species White-spotted Mouse-deer is present.

Behaviour/Social Organization Usually seen singly or in pairs.

Diet Feeds on grasses, herbaceous plants, tender leaves and fallen fruits including berries. My attempts to plant trees on a private nature reserve were met with a reduced success due to mouse-deer grazing them and nibbling the bark, which resulted in the death of plants. Mouse-deer probably play a role in thinning out emerging new forest plants under mother trees.

Subspecies None described. Recently split as a new species.

Size Head and body length 50.5–56 cm; tail 2.25–3 cm.
Weight 3.8 kg (female); 3.1 kg (male).
These measurements are based on Phillips (1980) and the range encompasses both species of mouse-deer.

Where to See Regularly reported from home gardens by people with homes around Bolgoda Lake. Sightings from home gardens around the Talangama Wetland will decline because of rapid habitat loss due to housing developments.

SPOTTED DEER *Cervus axis*

The Spotted Deer is the deer species most likely to be seen by visitors to national parks. Because of poaching, it is now virtually confined to protected areas. Spotted Deer are often found in association with Hanuman Langurs, with which they seem to have a symbiotic relationship. They probably assist each other with enhanced vigilance, and the Spotted Deer may gain some extra foraging benefit from the foliage broken by the langurs.

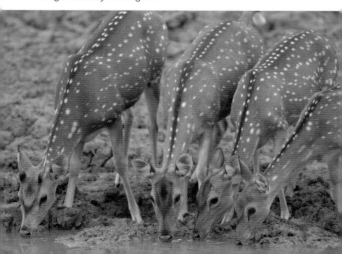

Habitat Grassy glades for grazing and forest cover for shade during the heat of the day. Found in scrub jungles of the dry lowlands.

Distribution Throughout the island (except the northern peninsula) in suitable dry-zone scrub jungles.

Behaviour/Social Organization Forms herds. A dominant male secures the mating rights to a herd of females. However, mixed herds are seen quite often, with several males mingling with females. Stags at times decorate their antlers with vegetation for displaying. Only the males develop antlers.

Diet Herbivorous. Spotted Deer will occasionally browse on low-hanging branches, but feed mainly by grazing on grasses.

Subspecies None.

Size Height at shoulder 90 cm.
Weight 85 kg.

Where to See National parks.

SAMBAR *Cervus unicolor*

The Sambar is the largest deer on the island. This is the 'elk' referred to by 19th-century European hunters in Sri Lanka; the true elk is not found in south Asia. The Sambar is easily identified by its uniform, dark chocolate brown coat and large size.

Sadly, the Sambar remains a popular quarry of poachers and is extremely wary, even within national parks and reserves. The one exception is in Horton Plains National Park. This is also the only place in Sri Lanka where the Sambar is seen in herds of any size. There are two reasons for this. One is the presence of large grasslands. The nutritional content of these grasslands is artificially high thanks to the escape from the nearby Ambewela cattle farm of an introduced species of grass. The second reason is the protection offered by the permanent presence of rangers and regular visitation by the public.

Habitat Sambar can occupy a wide range of habitats, from dense rainforests to open grasslands. They do, however, always need the cover of wooded thickets for lying up during the day.

Distribution Found throughout the island wherever sizeable pockets of forest and grassland remain. Populations in any significant numbers are increasingly being confined to the protected areas.

Behaviour/Social Organization Only the males of the species develop antlers. They battle for control of harems of females in a manner similar to that of males of other species of deer such as the Spotted Deer, *C. axis*. The main predator of the Sambar is the Leopard, although young may also be taken by Golden Jackals. In the dry lowlands, Sambar are wary of crocodiles when drinking. Even in Horton Plains National Park, where they have become habituated, they are active mostly at night. By evening, sizeable herds of up to 30 animals may gather in the open to forage on grass. In forested areas the stags may only associate with the females during the rutting season. Males often bellow to establish their territorial dominance. They also give a short, sharp bark when alarmed. This far-carrying bark often alerts other animals to the presence of a predator such as a Leopard.

Diet Herbivorous. Being a forest deer, the Sambar is mainly a browser, feeding on leaves and shoots of plants, but also grazing on grasses. In locations such as Horton Plains National Park, it benefits from having large grasslands for grazing as well as a rich cloud forest understorey, with plants such as Nillu, *Strobilanathes* spp., to browse.

Subspecies The subspecies *C. u. unicolor* is considered endemic to the island. It is not distinct in the field from subspecies on the Asian mainland.

Size Head and body length 1.51 m (female) to 2.55 m (male); height at shoulder 1.02 m (female) to 1.4 m (male).
Weight 129.5 kg (female); 304.5 kg (male).

Where to See Sightings are guaranteed in the evenings in Horton Plains National Park. Yala National Park probably gives the next best chance of seeing it. However, I have at times not glimpsed one for days.

INDIAN MUNTJAC or BARKING DEER *Muntiacus muntjak*

This is a uniformly brown-coloured, medium-sized deer with small antlers mounted on long pedicles. It is often seen singly, one prominent exception being Wilpattu National Park, where small groups can gather. Most of the time it is a thinly distributed animal. It has naturalized in many European countries, demonstrating its ability to adapt and to be able to live near people without giving itself away.

Habitat It likes areas where forest cover is interspersed with grasslands. The habitat at Wilpattu is ideal. Many of the highland patna grasslands with ridge-top forest also provide its preferred habitat.

Distribution Lives throughout the lowlands to the highlands, but is least abundant at the highest elevations. Nevertheless, at times in the early morning, small numbers can be seen grazing on the agricultural fields bordering the cloud forests close to Horton Plains National Park. Phillips (1980) states that its range in the south-east extends into Yala (Ruhunu) National Park, but I have never encountered it in Yala or come across a reliable account of its occurrence in recent years (I am not sure why, given that it is quite adaptable and Yala must provide it with good habitat similar to Wilpattu).

Behaviour/Social Organization Usually occur singly or in pairs, but – where there is grassland forest cover and they are not hunted – they will form small groups. The deer is hunted for its flesh and has become wary and nocturnal in most areas. The best clue to its presence is its barking call, usually uttered in alarm or when the bucks are in rut. The canine teeth are developed into tushes (small, tusk-like protrusions), which the deer use in defence and with which they can inflict damage on a predator.

Diet It forages on succulent grasses and tender leaves.

Subspecies The subspecies *M. m. malabricus* is confined to southern India and Sri Lanka.

Size Head and body length 88 cm (female); 1.22 m (male); height at shoulder 45.72 cm (female); 62.23 cm (male).
Weight 19.5 kg (female); 25.5 kg (male).

Where to See Wilpattu National Park is the best place.

WATER BUFFALO *Bubalus bubalis*

The Water Buffalo is not likely to be confused with any other mammal in Sri Lanka; it is the only member of the bovine family occurring in a wild state. There is debate as to whether wild buffalo is truly a native mammal or whether it originated from imported domestic animals that have gone wild for several hundred years. It is most likely that it is descended from imported animals, as the original range of wild Water Buffalo was in the Nepalese Terai, the plains of the Ganges and the Brahamaputra in Assam. It accounts for at least a few human deaths each year. Although these are large mammals, they can submerge themselves in water leaving only the inwards-curved horns and their nostrils above it. Covered in mud at a difficult angle, an adult bull may be overlooked. I have heard accounts of people passing a submerged bull only to have it emerge from the water and goring the walker from behind. In remote areas, people sometimes die from internal bleeding before they can be treated. Kithsiri Gunawardana, a good naturalist and keen butterfly photographer, was photographing butterflies when a buffalo attacked his companion. Fortunately, they avoided serious injuries. Treat all buffalos with caution and respect.

A popular Sri Lankan dessert is curd and honey made using the milk of the buffalo. The curd is similar in texture and colour to white yogurt.

Habitat Found in scrub forest with grassland and waterholes in the dry lowlands.

Distribution In the wild state it is almost entirely restricted to the national parks in the dry lowlands.

Behaviour/Social Organization Buffalos have a fairly close-knit structure. When a herd is moving, the hierarchy and the sex of an individual determine its position in the herd. The dominant male may at times follow behind another adult male, which takes the pathfinder position. When several buffalo clans coalesce into a large herd, the clan leaders take the pathfinder positions on the edges of the herd when the herd is moving. Females, especially cows with young, are positioned in the centre. Sexually mature bulls often leave the herds and may join bachelor herds or remain as loners.

Buffalos do not generally defend a territory as such, but they do have a home range in which they move around. They are not very receptive to individuals joining a clan, although clans themselves coalesce into larger herds. Adult bulls are most likely to be seen in herds with females and young when adult females are in oestrus.

I was once on a morning game drive in Yala when we came across an adult female bellowing in great anger and distress. A few hundred metres away lay her calf, probably killed during the night by a Leopard.

We waited at a distance, but within view of the calf. The mother ensured that the Leopard did not approach the calf. Several hours later, the entire herd approached the site where the calf lay. One by one, the herd members sniffed the calf. It was easy to be anthropomorphic and imagine they were paying their last respects to the dead calf before the mother followed the rest of the herd, leaving the calf to the hunter.

The Water Buffalo, Wild Pig and Sambar are the three mammals commonly found wallowing in mud. Of these three, only the buffalo is fond of immersing itself in the water, which it does for great lengths of time.

Diet Browses on aquatic plants. Grazes short grasslands.

Subspecies The subspecies *B. b. bubalis* is found in India and Sri Lanka.

Size Head and body length 2.7 m; height at shoulder 1.5 m.
Weight Almost 909 kg.

Where to See 'Wild' Water Buffalos can be seen easily at the waterholes of the dry-zone national parks such as Yala, Wasgomuwa and Uda Walawe. They are indistinguishable from many of the domesticated animals seen ploughing paddy fields.

INDIAN PANGOLIN *Manis crassicaudata*

The Pangolin is an unmistakable mammal with a small, long head and scales on its upper body. The scales provide defence from predators when it curls up to protect its soft underparts. The hard, sharp-edged scales can also inflict injuries on predators.

Habitat Lives in a mix of forest habitats from dry thorn scrub to wet-zone forests, where its main food – termites and ants – is found. The flesh of the Pangolin is considered tasty and it has been eliminated in areas where it strays into contact with people.

Distribution Distributed up to the mid-hills to about 1,050 m. Its distribution coincides with the occurrence of its main prey, the termite.

Behaviour/Social Organization No formal studies have been made of this animal in the wild in Sri Lanka and very little is known about its social organization. This seems to be a naturally scarce mammal; even in the national parks where it is protected it is rarely encountered at night. But this could be more of a reflection of the fact that that very little research work is undertaken on nocturnal mammals. Outside protected areas numbers may be very low due to it being hunted by people for its flesh. Most of my records of it are by coming across the remains of a dead Pangolin, although once I had the good fortune to watch a Leopard cub of around 12 months playing with a Pangolin. It feigned death until the cub got bored and left. The Pangolin then got up and shuffled away. Pangolins walk on the sides of their forefeet to reduce wear and tear on their digging claws, which are strong enough to rip open anthills. They also climb trees to devour the ants from their nests.

Diet It feeds mainly on termites (white ants) and ants.

Subspecies None.

Size Body length 48–81.7 cm; tail 40.6–59.4 cm.
Weight 9.1 kg (female); 19.1 kg (male).

Where to See No reliable site. On night safaris on roads through forests.

Dry lowland race of Grizzled Indian Squirrel

Palm Squirrel (F. p. brodei)

Squirrels are in the family Scuiridae in the order Rodentia. This order encompasses many other mammals that behaviourally and in terms of appearances may not seem to be closely related. Anatomical features such as dentition link rodents, although many people do not view squirrels and porcupines as rodents since that term is most commonly associated with rats and mice. The porcupine is in the subfamily Hytericidae and the rats and mice are in the family Muridae.

The family Scuiridae comprises around 200 species that live worldwide, except in the Arctic, Antarctic and Australia. Four diurnal squirrel species and two nocturnal species of flying squirrel are found in Sri Lanka. All four of the diurnal squirrels are in the genus *Funambulus* and are marked with three stripes on their back. A long, furry tail characterizes these squirrels. Vocalizations form an important part of their behavioural strategy. They have far-carrying alarm calls and also maintain territories by calling loud from a high vantage point. All of the Sri Lankan diurnal squirrels have a tendency to follow flocks of birds. The sexes of the squirrels do not show pronounced differences in size.

LAYARD'S STRIPED SQUIRREL *Funambulus layardi*

A dark squirrel about the size of the common Palm Squirrel, Layard's Striped Squirrel often betrays its presence with a bird-like, chittering call (it is not quite as vocal as the Palm Squirrel). Good views reveal the beautiful 'flame-stripe' on its back. It has three dorsal stripes, with the middle one being the widest and longest.

Habitat Keeps to tall forests in the wet zone. It will visit home gardens adjoining good-quality forest, but is absent in areas where virgin forest is absent.

Distribution From the lowlands to the higher hills up to around 1,220 m. In the higher hills, it is not as abundant as the small Dusky Squirrel and is absent at the highest elevations. The species is confined to southern India and Sri Lanka.

Behaviour/Social Organization Usually seen singly or in pairs. It nests in a drey like the Palm Squirrel. Its behaviour has many parallels with the smaller Dusky Squirrel. Both species often follow mixed-species bird flocks in forests such as Sinharaja and Morapitiya. They are arboreal and will come to the ground only where they have to cross a path to climb the trees on the other side. Over the years they have become tolerant of visitors at sites such as Sinharaja, but remain wary and will take flight at the slightest hint of danger.

Diet Mainly tender leaves, fruits, nuts and lichens. It will also take insects and grubs.

Subspecies Two subspecies have been described. The Flame-striped Squirrel, *F. l. layardi*, is considered to range from the East Matale hills to Kithulgala and Adam's Peak. The Western Flame-striped Squirrel, *F. l. signatus*, lives in forests around Ratnapura to Galle.

F. l. signatus

The distinction between the two subspecies is considered to be the median dorsal stripe: yellowish buff for *F. l. layardi*, flame-coloured in *F. l. signatus*. The two subspecies do not occupy distinct climatic zones and animals can show some variation in the colour of their stripes, so the subspecies distinction is doubtful.

Size Head and body length 15.5–16.8 cm; tail 14.2–14.9 cm.
Weight 143 g.

Where to See The bird tables at the research station in Sinharaja at Martin's, the barrier gate and the research station. Also look for squirrels in mixed-species feeding flocks of birds.

F. l. signatus

PALM SQUIRREL *Funambulus palmarum*

This, the most common Sri Lankan squirrel, makes its home in cities, home gardens and forests.

Habitat Although it can be seen in urban environments, the Palm Squirrel needs tall trees in which it can roost for the night or shelter during the heat of the day.

Distribution The four subspecies are spread throughout the island to the highlands.

Behaviour/Social Organization Palm Squirrels have an interesting symbiotic relationship with Yellow-billed Babblers. Every flock of

F. p. favonicus

Above and below: F. p. favonicus

foraging babblers seems to have a group of Palm Squirrels. Different pairs of squirrels holding territories in different parts will join a flock of roving babblers. This may give the impression that a flock ties up with one pair of squirrels when several pairs of squirrels may be involved. Unusually for mammals, Palm Squirrels spend a great part of their time vocalizing. This behaviour is more typical of birds. However, unlike in birds, the squirrels' repertoire is limited to a monotonously repeated, high-pitched call. This is probably the most vocal mammal in Sri Lanka.

Diet Mainly plant matter such as shoots, leaves, seeds and fruits. Occasionally takes insects and grubs.

Subspecies The four subspecies of Palm Squirrel are separated by the colour of the dorsal stripes. Due to individual variation and the effects of lighting, separating them in the field is not straightforward. The geographical distribution is an added clue, but where subspecies intergrade in contact zones, subspecific identification is difficult.

In *F. p. favonicus* all the dorsal stripes are brown-buff. It lives in the west. The median dorsal stripe in *F. p. olympius* is white; other stripes are whitish-buff. It lives in the highlands. The general hue of *F. p. kelaarti* is light grizzled grey; all the dorsal stripes are white or almost white. It lives in the south. In *F. p. brodei* the general hue is light grizzled grey; all the dorsal stripes are whitish-buff. It lives in the north.

Size Head and body length 14–16 cm.
Weight 115 g.

Where to See Home gardens.

DUSKY STRIPED SQUIRREL *Funambulus sublineatus*

This is a small, dark squirrel that can look even darker in the dim recesses of a rainforest. There are three faint dorsal stripes on its chocolate-brown coat. The tail is usually not as long as the body – this species is relatively short-tailed compared with the other squirrels on the island. On first impressions it seems a bit misshapen because of its short tail.

Habitat Frequents good-quality forest patches remaining in the wet zone. Tends to feed mostly on the lower to mid-layer of trees, often going up and down trees; fond of scampering along fallen logs and exploring crevices.

Distribution Found in the wet zone from the lowland to the highlands where good-quality forests remain. In Nuwara Eliya it can sometimes be seen in Victoria Park, but only because Nuwara Eliya is surrounded by good-quality cloud forests in the Mount Pedro forest reserve. Although it may visit home gardens, is has not adapted to a totally urban habitat like the Palm Squirrel.

Behaviour/Social Organization Occasionally runs across a footpath to gain access to trees on the other side, but rarely feeds on the ground as the Palm Squirrel does at times. A mixed-species feeding flock of birds in Sinharaja will almost always be accompanied by one of these dainty squirrels. They are frequently seen in pairs. Their contact call is bird-like. A squirrel will go about its business provided an observer stays still, but it is a wary animal. In over 25 years of visiting Sinharaja rainforest, I have observed how in recent years it has taken to feeding at bird tables, but remains wary and ready to flee at the slightest inkling of danger.

Diet A mixed diet of flowers, tender shoots and fruits, as well as invertebrates such as insects and grubs.

Subspecies One subspecies, *F. p. obscurus*, in Sri Lanka. The Dusky Squirrel is endemic to southern India and Sri Lanka.

Size Head and body length 11.2–11.6 cm.
Weight 75 g.

Where to See Bird tables at Sinharaja at Martin's, near to the barrier gate and at Martin's. Also seen following mixed-species feeding flocks of birds and easy to see in the cloud forests in and around Horton Plains National Park.

The Giant Flying Squirrel is a spectacular animal to watch when it engages in a long glide. Unfortunately, because of its nocturnal habits most people do not see one. Its upperparts are similar to those of the black and yellow race of the Giant Squirrel, but the underparts are pale not yellow. Overall, it lends the appearance of a large, blackish squirrel. The smaller Ceylon Small Flying Squirrel, *Petinomys fuscocapillus layardi*, is also pale on the underparts, but is rufous-chestnut on the upperparts.

Habitat This arboreal squirrel seems to be most common in disturbed habitats, where tall trees are interspersed with open glades. As it spends the day and raises its young inside a tree hollow, the presence of old trees is probably a limiting factor for providing daytime sleeping sites.

Distribution Mostly in the mid-hills, but also in highlands to Horton Plains.

Behaviour/Social Organization Very little is known about the social organization of this nocturnal squirrel. At sites such as the Hotel Tree of Life, a number of individuals seem to concentrate in a small area. I have observed several feeding on a fruiting tree without any fighting taking place. They have membranes on the sides of their body

joined to the forelimbs and hind limbs. These are splayed out while gliding. I once got a fright when a squirrel I was watching leapt off a branch and headed straight towards my head. As it came overhead it seemed to stretch the membrane out fully to maximize the lift and actually gained height and landed on the trunk of a tree away from me. I estimate that the squirrel glided for over 30 m. I have been told stories of the squirrels gliding across rivers between banks.

Diet The bulk of their food seems to be fruits, but they also eat tender shoots, bark and invertebrates. They will visit home gardens to feed on mango and tamarinds.

Subspecies The subspecies *P. p. lanka* described from Sri Lanka is unique to the island.

Size Head and body length 43.2–47 cm; tail 49.5–54.6 cm.
Weight 2–2.3 kg.

Where to See The Hotel Tree of Life at Yahalathenna near Kandy has a colony of these squirrels, which emerge at dusk. If you are observing them, avoid using harsh spotlights. See the section on watching animals at night (page 7).

GRIZZLED INDIAN or GIANT SQUIRREL *Ratufa macroura*

R. m. macroura

This is the largest of the four species of squirrel in Sri Lanka. It becomes habituated to people and will readily approach them at picnic spots and hotels to take fruits.

Habitat Requires forests with tall, mature stands.

Distribution Three subspecies are spread throughout the island, to the highest mountains.

Behaviour/Social Organization Usually found in pairs that maintain a territory. It has a loud, almost hysterical call, which it uses to warn of danger.

Diet A wide range of plant matter, from roots to leaves and fruits.

Subspecies The black and yellow highland race *R. m. macroura* is similar to the race in the wet lowlands, but the tail is frosted white. In the wet lowlands race *R. m. melanochra* the tail is completely black. The race *R. m. dandolena* in the dry lowlands has distinctive brown and buff fur.

Size Head and body length 33–40 cm; tail 35–41 cm (the female a little bigger).
Weight 1.7 kg.

Where to See The race *R. m. dandolena* can be seen easily in the national park campsites. *R. m. melanochra* lives in wet-zone rainforests such as Sinharaja, while *R. m. macroura* is in Horton Plains National Park.

Above: R. m. macroura. Below: R. m. melanochra

INDIAN GERBIL *Tatera indica*

Besides the Black or House Rat, *Rattus rattus,* and Brown Rat, *R. norvegicus,* this rat is the other species most likely to be seen by visitors on a wildlife tour. This is because of its habit of foraging on forested tracks after nightfall. It looks quite pale in the glare of headlights. It is easily identified by its long, fur-covered tail, which has a tuft of black fur at the end. It has a bounding run that is different from the scurrying run of *Rattus* rats, from which its alternative common name, Antelope-rat, arises.

Habitat It is found in a range of habitats from dry-zone jungle to wet-zone rainforest. It also occupies plantations (a severely degraded habitat). The Indian Gerbil does not seem able to compete with other species of rat and mice in urban environments with a high density of people.

Distribution Found throughout the island up to the mid-hills. It is probably more common, if not more easily seen, in the scrub-forested areas of the dry lowlands.

Behaviour/Social Organization Gerbils are aggressive animals and are known to attack and kill animals larger than themselves. Both males and females dig their own burrows. The females typically build two bolt-holes. The burrows are quite often built on open ground without cover, but occasionally they make their burrow within a termite hill. They do not habitually store food.

Diet Omnivorous. Gerbils will not only eat a range of plant matter from grain to the seeds of the introduced rubber tree, but will also kill and eat small mammals and reptiles. Eggs, grubs and insects are all eaten.

Subspecies The subspecies *T. i. ceylonica* is endemic to the island.

Size Head and body length 16–20.5 cm; tail 20.5–21.7 cm.
Weight 148 g.

Where to See Gerbils are frequently encountered when driving at night on forested roads in the dry zone. The roads outside Yala

National Park throw up a few if you are arriving late to a hotel such as the Yala Village Hotel. The last few kilometres of forested road on the unsurfaced track to the Yala Village Hotel is good for a number of nocturnal mammals. See also the guidelines on watching nocturnal mammals without adversely affecting them (page 7).

GREATER BANDICOOT-RAT *Bandicota indica*

This is the largest of the rat-like rodents, reaching a head and body length of 17.5 cm – with the tail it is just under 30 cm long. Despite their rather fierce appearance, bandicoot-rats are not aggressive. The name bandicoot is derived from the Tamil word *pandicuttie*, used in India to mean young pig, a reference to the pig-like grunts uttered when the animal is frightened. The fur is coarse and the tail is naked and scaly. The ears are large with a longish head. The young of this species may be confused with the Lesser Bandicoot-rat, *B. bengalensis*. Adult Greater Bandicoot-rats have relatively large feet in relation to body and head size.

Habitat Although this rat is found around human habitation, it does not habitually venture indoors as House and Brown Rats do. Its diet is mainly vegetarian, with paddy being especially favoured. It often frequents areas near paddy fields, although every patch of scrub forest will also hold bandicoot-rats. When paddy fields are flooded for cultivation, bandicoot-rats dig a fresh set of burrows in adjoining higher ground. The animals favour areas near water for building their burrows and are good swimmers.

Distribution Found throughout the island but seems to be less common in the dry zone.

Behaviour/Social Organization Bandicoot-rats build long galleries running along the embankments of paddy fields. Side tunnels are employed to deposit earth away from the main gallery. The side tunnels are generally kept plugged loosely with earth and serve as emergency exits in case a predator such as the Rat Snake enters the burrow. The main gallery has at least one circular chamber in which young are raised

and a few storage chambers where grain is stored. The rats store a prodigious amount of grain, which is at times retrieved by digging out the burrows. Phillips also records that in the Deccan in India, a caste of people used to capture bandicoot-rats for food. Greater Bandicoot-rats are mostly nocturnal, although I have been startled a few times in Colombo by a daytime crop raider visiting my garden.

Diet Vegetarian, eating paddy and tuberous roots of plants such as yams and potatoes. It is considered one of the biggest pests for cultivators of paddy.

Subspecies Two subspecies are described from Sri Lanka: *B. i. gracilis* and *B. i. insularis*. The latter is found in the area around Jaffna and is more uniformly pale on the abdomen; the head and neck have larger, shorter and sparser fur. The upperparts are a sandy brown and paler than those of the subspecies *B. i. gracilis*.

Size Head and body length 17.5 cm.
Weight 0.8–1.1 kg.

Where to See Village gardens and paddy fields at night.

HOUSE RAT *Rattus rattus*

This is the largest of the four or five species of rat in Sri Lanka, and there is some confusion in the taxonomy of these species in the country. Until recently it was believed that the largest rat (with a head and body length of about 8–10 inches) found in Sri Lanka was the Brown Rat, *Rattus norvegicus*. House Rats, *Rattus rattus*, were considered smaller, with a body length of 6–7 inches. However, on-going work by researchers of the Wildlife Heritage Trust suggests that the Brown Rat may not in fact occur in Sri Lanka.

The larger rats in Sri Lanka have tails that are stout at the base, and the tail length is less than the length of the body and head. They also have blunt muzzles, finely furred short ears, large feet and scaly tails, and are characterized by a stout, heavy build and shaggy fur. In contrast, what is traditionally considered as a House Rat has a slim tail that is longer than the head and body length. More work is currently being done to establish whether these are different populations of the House Rat or actually different species.

Habitat House Rats prefer areas located near human habitation. The larger rats whose identity is yet to be ascertained prefer areas near water. House Rats thrive in and around human habitation because of the availability of food from household waste.

Distribution Found throughout the island.

Behaviour/Social Organization Studies on the similar Brown Rats have revealed that they have a peculiar mating system in which several males mate with a female without any apparent competition between them. However, a male that is mating subsequently mates

more times with a female. This may be a strategy to improve its chances of being the one to inseminate the female. The females build a nest in a hole lined with straw, dead leaves, paper and so on. The young of rats and mice may use ultrasonic squeaks to communicate with their mother. Rats may only live to up to two years of age, but a female is sexually mature by six months and may raise as many as four or five broods each year. The rats seen scrambling up walls and along rafters are almost certainly House Rats.

Diet Highly omnivorous, eating a variety of animal and vegetable matter. Householders know only too well that even cakes of soap are eaten. Studies on the Brown Rat have shown that it will leave food untouched if it crops up somewhere unexpected. This is called neophobia (fear of the new), and is a survival strategy that no doubt protects the rats from numerous attempts by man to poison them. However, if a rat detects the smell of the new food on another rat that is alive and healthy, it will consume the new food. Members of a rat community are constantly learning about new sources and new items that they eat by smelling each other's mouths to gauge what has been eaten. House Rats probably share these survival strategies that have been studied in Brown Rats in Europe.

Subspecies Five subspecies have been described from the island: *R. r. rattus*, *R. r. alexandrinus*, *R. r. rufescens*, *R. r. kandianus* and *R. r. kelaarti*. The subspecies *R. r. rattus* and *R. r. alexandrinus* are considered to comprise populations whose origins lie near ports, having been accidentally brought in by ships. *R. r. rufescens* is thought to be a native of the dry zone, *R. r. kandianus* an endemic subspecies of the mid-hills, replaced by the endemic subspecies *R. r. kelaarti* in the highlands. Even at a single site, there can be a high degree of variation among individuals.

Size Head and body length 15–24 cm; tail 17–25 cm.
Weight 92–106 g.

Where to See Open garbage sites frequently have communities of rats. They are wary of predators such as cats and crows. Garbage pits with cover around them receive frequent daytime visits.

115

CRESTED PORCUPINE *Hystrix indica*

The Crested Porcupine is a relatively common mammal that is seldom seen because of its strictly nocturnal behaviour. Even in protected areas, where it is safe from poaching, it chooses to emerge at night. The usual clue to its presence is a discarded quill.

The Crested Porcupine is blackish brown and white, clothed with pointed defensive quills. The quills on the tail form a train. When a quill enters an animal's flesh it is likely to work its way further in because of the texture of the quills. A Leopard or Tiger that gets a quill in its paw may find that the shaft works its way in and leaves it with a debilitating injury. In India many man-eaters have their roots in an encounter with a porcupine. In Sri Lanka, Leopards probably leave porcupines alone as they have a good supply of alternative prey.

Habitat Crested Porcupines frequent wooded habitats and have readily adapted to home gardens. They are powerful diggers and will excavate their own sleeping burrow. Thus the availability of sleeping sites does not become a population-limiting factor. Their broad diet may explain their continued survival in urban habitats.

Distribution Found throughout the lowlands to the highlands, but less common at the higher elevations.

Behaviour/Social Organization Usually seen in pairs. Sometimes several individuals may sleep together. They are very resilient to human presence. I know of a pair on a privately owned nature reserve that abandoned a sleeping burrow while a pond was being excavated. They moved back a few days later when the digging had been completed.

Diet A wide range of plant matter from roots and tender leaves to fruits. They feed on fallen fruits from trees such as Jak, *Artocarpus heterophyllus*, Goraka, *Garcinia queasita*, and Cashewnut, *Anarcardium occidentale*, which are popular in home gardens. They also enjoy raiding the yams of plants such as the introduced Manioc, *Manihot esculenta*, in home gardens, a habit that does not endear them to people.

Subspecies None described. The species has a wide distribution from southern Arabia through India to Sri Lanka.

Size Males are biggest, with head and body length 66–67 cm; tail 13–17.85 cm.
Weight 11.8–14.5 kg.

Where to See Home gardens of people are the best sites. They are regularly seen around locations such as Talangama, Kotte and Bolgoda Lakes; quite densely populated areas that retain patches of secondary growth.

BLACK-NAPED OR INDIAN HARE *Lepus nigricollis*

This is the only species of wild hare or rabbit in Sri Lanka. Surprisingly, it still holds out on the outskirts of Colombo. Outside the protected areas it is nocturnal due to poaching and threats from domestic dogs and cats.

Juvenile Black-naped Hares

Habitat Found in forested areas interspersed with glades of grassland.

Distribution Throughout the island to the highest mountains.

Behaviour/Social Organization During the day the Black-naped Hare prefers to lie up in a form (a patch of vegetation in which it snuggles and is well camouflaged). Where it is protected, such as in national parks, it may be active during the day, or at least in the early mornings. Otherwise it is seen when it emerges to roadside verges after dark.

Diet Herbivorous. Grasses, shoots, young leaves and so on.

Subspecies The race *L. n. sinhala* is unique to Sri Lanka.

Size Head and body length 45–50 cm; tail 7–10 cm.
Weight 2–3 kg.

Where to See Yala National Park.

GLOSSARY

Adult Sexually mature.

Arboreal Tending to inhabit the canopy of trees.

Canines Pointed, generally long teeth behind the incisors.

Carnivore Meat-eating animal.

Crepuscular Active at dawn and dusk.

Diurnal Active during the day.

Dorsal On the upperparts of the body. On mammals, for example, dorsal stripes refer to stripes running along the back.

Endemic Native to a particular region or restricted area.

Feral animal Domestic animal that has escaped from captivity and breeds in the wild.

Foraging Searching for food.

Herbivore Any animal that feeds entirely on plants.

Home range Area in which an animal usually lives and carries out its day-to-day activities.

Incisors Sharp-edged front teeth, usually in both the upper and lower jaws.

Infant Baby animal that is entirely dependent on its mother.

Insectivore A mammal that feeds mainly on insects.

Juvenile Young animal that is not yet fully grown, but partially or completely dependent on its mother.

Melanism Unusual dark colouring due to the presence of the pigment melanin.

Migration Movement, usually seasonal, from one region or climate to another for the purposes of feeding or breeding.

Muzzle The part of an animal's face in front of its eyes, including the nose.

Nape The back of the neck.

Nocturnal Active at night.

Omnivore Any animal that preys on other animals for its food.

Opportunist Flexible behaviour consisting of exploiting circumstances to take a wide range of food items.

Pedicles Bony growths on the skull that support the growth of antlers.

Posterior Towards the back.

Predator Animal that forages for live prey.

Riparian In close association with rivers and river-bank habitats.

Rostrum The narrower part of the skull in front of the eyes.

Rut Period when deer are in season and the males challenge other males for dominance to gain access to females.

Sounder Collective name given to pigs.

Terrestial Active on the ground.

Territory A restricted area inhabited by an animal, often for breeding purposes, and defended against other individuals of the same species.

Tines The prongs on deers' antlers.

Underparts A mammal's throat, chest, abdomen and insides of the legs.

Ungulate Mammal that has its feet modified as hoofs.

Upperparts A mammal's top part or back, including the outsides of the legs.

Ventral On the underside (not to be confused with the vent, which is the area around the anus).

INFORMATION FOR VISITORS

Some preparation can make a lot of difference in an accessible small island like Sri Lanka. A good deal of information is available on websites and in traditional books.

Time to Travel

Because of the climatic variation there is usually some part of the country that is dry and enjoying good weather at any time of year. Generally speaking Sri Lanka is a year-round destination for wildlife enthusiasts and photographers.

The period from January to April marks the warm, dry season in the lowlands. This is the favoured time for many visitors seeking a break from the northern hemisphere winter. However, the highlands at Nuwara Eliya and Horton Plains, for example, may experience frost in January and February.

The inter-monsoonal lull is between February and April. February is the driest month in the south-west. The best time for visiting birdwatchers extends from November to April. February is a good month as it is largely free of rain. Migrant birds visit during this period and add to the species tally. Wildlife viewing in the national parks in the dry zone is best during dry, hot and dusty times, when streams are reduced to a mere trickle and animals are concentrated around waterholes.

Information on the Internet

The Internet flyway in the Oriental Bird Club's website (www.orientalbirdclub.org) is a good place to start a search for recent trip reports (these are mostly relevant to birds but also include mammals). Email discussion forums such as the one on UK Bird Net and NatHistory South India are other good places from which to gain information. Past copies of the Sri Lanka Wildlife eNewsletter compiled by the author Gehan de Silva Wijeyeratne are on the Oriental Bird Club website and www.jetwingeco.com. There are over 1,000 pages of content on www.jetwingeco.com including reports on watching and photographing mammals.

Useful Books for Visitors

A pocket photographic guide such as this one is adequate for seeing most mammals that a visitor may encounter on a short trip. It also includes all species a resident with a casual interest is likely to see. Serious mammal enthusiasts should look for *A Manual of the Mammals of Sri Lanka* by W.W.A. Philips.

Invaluable publications for pre-trip reading and use in the field include *A Birdwatcher's Guide to Sri Lanka*, which is published by the Oriental Bird Club and lists 21 key sites, and *Sri Lankan Wildlife* by Bradt Travel Guides, which is lead authored by Gehan de Silva Wijeyeratne. *Sri Lanka National Parks and Reserves* and a number of photographic guides to butterflies, dragonflies, mammals and other items of interest are available free as downloadable PDF-format files on www.jetwingeco.com.

Organized Tours

A number of overseas companies operate tours to Sri Lanka. These may not always be convenient for a family with young children or for those for whom mammals and other natural history may be a priority over birds. A number of local tour companies can organize a package involving mammals and natural history, culture and even chilling out on the beach, all tailored to individual requests. I have listed some with which I am personally acquainted. Some are long-standing companies that offer wildlife tours. Others are relatively new companies with a strong focus on natural history. Yet others are specialist subsidiaries of the big names in the travel industry. Many are based in Colombo (the international phone code for Colombo is + 00 + 94 +11). If you book a tour with a company licensed by the Sri Lanka Tourist Board that uses licensed, experienced guides, things will go smoothly. Some tour operators are also members of the Sri Lanka Inbound Tour Operators Association (SLITO).

A. Baur & Co (Travel) Ltd
Travel.baurs.com

High Elms Travel (Pvt) Ltd
www.highelmstravel.com

Aitken Spence Travels
www.aitkenspencetravels.com

Lanka Sportreizen
www.lsr-srilanka.com

Bird and Wildlife Team (Pvt) Ltd
www.birdandwildlifeteam.com

Jetwing Eco Holidays
www.jetwingeco.com

Birdwing Nature Holidays
www.birdwingnature.com

Quickshaws Tours
www.quickshaws.com

Eco Team (Pvt) Ltd
www.srilankaecotourism.com

Walk with Jith
www.walkwithjith.com

Hemtours
www.hemtours.com

Clubs and Societies

Sri Lanka Natural History Society.
Website: www.slnhs.lk
Founded in 1912, the Sri Lanka Natural History Society (SLNHS) has remained an active, albeit small society with a core membership of enthusiasts and professionals in nature conservation. It organizes a varied programme of lectures and slide presentations for its members. The subject matter embraces all fields of natural history, including marine life, birds, environmental issues and the recording thereof via photography and other techniques. The society organizes field excursions regularly, including day trips as well as longer excursions with one or more overnight stays.

Field Ornithology Group of Sri Lanka (FOGSL),
Email: fogsl@sit.lk
Website: www.fogsl.net
FOGSL, the Sri Lankan representative of BirdLife International, is pursuing the goal of becoming a leading local organization for bird

study, bird conservation and carrying the conservation message to the public. It has a programme of site visits and lectures throughout the year, and also publishes the Malkoha newsletter and other occasional publications. Education is an important activity, and FOGSL uses school visits, exhibitions, workshops and conferences on bird study and conservation to promote its aims.

Wildlife and Nature Protection Society (WNPS),
Email: wnps@sltnet.lk
Website: www.wnpssl.org
The WNPS publishes a biannual journal, *Loris* (in English) and *Warana* (in Sinhalese). *Loris* carries a wide range of articles, ranging from very casual, chatty pieces, to poetry and technical articles. The society also has a reasonably stocked library on ecology and natural history. Various publications, including past copies of *Loris*, are on sale at its office.

Young Zoologists' Association of Sri Lanka,
Email: srilankaza@gmail.com
Website: www.yzasrilanka.lk
At present, the YZA has nearly 100 school branches and has also set up branch associations. The bulk of its membership comprises schoolchildren and undergraduates, the rest being graduates, professionals and nature enthusiasts.

Ruk Rakaganno (Tree Society of Sri Lanka),
Email: rukraks@sltnet.lk
Ruk Raks aims to combat the destruction of Sri Lanka's forests. It conducts urban and rural tree-planting programmes and engages in activities to raise appreciation of trees, particularly among the young. Current activities include replanting and awareness programmes in coastal areas, maintenance of a nursery of primarily indigenous trees and management of the IFS—Popham Arboretum in Dambulla. The Tree Society organizes both seminars and field trips.

Sri Lanka Wildlife News
Sri Lanka Wildlife News is a quarterly compilation of news, events, birds, Leopard and other wildlife sightings, trip reports, articles and recent publications of interest to wildlife enthusiasts, conservationists and photographers. To receive this free email-based newsletter, email gehan@jetwing.lk with 'subscribe wildlife news' in the message header.

REFERENCES

Key Sources

Manual of the Mammals of Sri Lanka by W.W.A. Phillips remains an outstanding piece of work. In many respects, it will always remain an essential source of reference. It was a key source of reference for taxonomic information on the various levels of classification in this book (orders, families, genera, subspecies and so on).

For details on size and weight in the current volume, I referred to Phillips as well as to *A Field Guide to Indian Mammals* by Vivek Menon. Gaps on weight were filled in for me by Rajith Dissanaike, Rohan Pethiyagoda and Suyama Meegaskumbura. Some of the weights are based on information on just a single individual, illustrating the amount of work that still remains to be done on Sri Lankan mammals.

The taxonomic arrangements vary from one author to another. I followed the order in an unpublished checklist given to me by Priyantha Wijesinghe updated for changes in endemic status and scientific names based on Weerakoon and Goonetilake (2006).

Information on behaviour was based on field observations, Phillips' *Manual of the Mammals of Sri Lanka* and also Richard Despard Estes' *The Behaviour Guide to African Mammals* (many Asian species are shared with Africa).

The species accounts and accounts at higher levels for genera and families compiled by Richard Despard Estes provided valuable insights into the behaviour of Asian species. This fantastic book is in my view an essential book in the library of any serious mammal watcher. For behaviour of mammals in general and for behavioural insights of species shared between Europe and Asia, David MacDonald's *European Mammals: Evolution and Behaviour* is excellent.

Apps, P. and du Toit, R. (2000), *Creatures of Habit: Understanding African Animal Behaviour*, text by Peter Apps, photography by Richard du Toit, Struik Publishers, Cape Town, p. 160.

Bernede, L. and Gamage, S. (2006), *A Guide to the Slender Lorises of Sri Lanka*, Primate Conservation Society of Sri Lanka, Colombo, p. 8.

Estes, R.D. (1991), *The Behaviour Guide to African Mammals: including Hoofed Mammals, Carnivores, Primates*, Russel Friedman Books, South Africa.

Groves, C.P. and Meijaard, E. (2005), 'Interspecific variation in Moschiola, the Indian chevrotain', in 'Contributions to biodiversity exploration in Sri Lanka', *The Raffles Bulletin of Zoology 2005*, Supplement No. 12, National University of Singapore, pp. 413–21.

Macdonald, D. (1995), *European Mammals: Evolution and Behaviour*, HarperCollins, London.

Menon, V. (2003), *A Field Guide to Indian Mammals*, Dorling Kindersley (India), Delhi.

Phillips, W.W.A. (1980), *Manual of the Mammals of Sri Lanka*, 2nd revised edition, Wildlife and Nature Protection Society of Sri Lanka, Colombo.

Weerakoon, D.K. and de A. Goonetilake, WLDPTS (2006), 'Taxonomic status of the mammals of Sri Lanka', in (ed)

Bambaradeniya, C.N.B., *The Fauna of Sri Lanka: Status of Taxonomy, Research and Conservation*, The World Conservation Union, Colombo, Sri Lanka & Government of Sri Lanka, pp. 308 + viii.

Recommended Books

Bedjanic, Matjaz, de Silva Wijeyeratne, G. and Conniff, K. (2007), *Dragonflies of Sri Lanka*, Gehan's Photo Guide Series, Jetwing Eco Holidays, Colombo.

d'Abrera, B. (1998), *The Butterflies of Ceylon*, Wildlife Heritage Trust, Colombo.

de Silva Wijeyeratne, G., Warakagoda, D. and de Zylva, Dr T.S.U. (2000), *A Photographic Guide to the Birds of Sri Lanka*, New Holland, London.

de Silva Wijeyeratne, G. (2005), *Sri Lanka National Parks and Reserves*, Jetwing Eco Holidays, Colombo.

de Silva Wijeyeratne, G. (2006), *Wildlife of the Dry Lowlands, a Photographic Guide to the Commoner Animals and Plants of the Dry Lowlands*, Gehan's Photo Guide Series, Jetwing Eco Holidays, Colombo.

de Silva Wijeyeratne, G. (2007), *Sri Lankan Wildlife*, Bradt Travel Guides, UK.

Jayawardene, J. (2004), *The Elephant in Sri Lanka*, Colombo, Sri Lanka.

Pethiyagoda, R. (1998), *Ours to Protect Sri Lanka's Biodiversity Heritage*, WHT Publications (PVT) Limited, Colombo.

PHOTOGRAPHIC NOTES

Although the majority of the species described in this book were photographed in the wild, some of the images are of animals in captivity.

The majority of my still photography is now taken with a Canon EOS 1DS Mark II 16.7 megapixel full-frame sensor digital SLR. A Canon EOS 10D is my second body. I use two film SLR cameras as additional bodies when needed. For film, I use FujiFilm transparency film. I only use Canon extra-low dispersion optics designed for professional use.

For big game safaris and bird photography, a Canon 600 mm f4 is my favourite lens. When extensive walking is involved, I use a Canon 300 mm f4 or a Canon 100-400 mm IS lens, at times coupled with a Canon 1.4x converter. For macro work I use Canon's professional 100 mm f2.8 macro lens.

PICTURE CREDITS

Principle photography by Gehan de Silva Wijeyeratne.
Additional photography:
Lilia Bernede 23 (top), 24, 27 (bottom), 29–30 (Red Slender Loris).
Nevile Buck 9, 55, 63 (bottom), 64 (Fishing Cat); 58, 61 (Jungle Cat); 59, 62 (Rusty-spotted Cat).
Anna Nekaris 27 (top), 31–2 (Grey Slender Loris).
Suyama Meegaskumbura 13–14 (House Shrew); 112 (Indian Gerbil); 115 (House Rat).
Ruchira Somaweera 88 (White-spotted Mouse-deer); 113 (Greater Bandicoot-rat); 116–7 (Crested Porcupine).
Wipula Yaka 17 (Short-nosed Fruit Bat).

ACKNOWLEDGEMENTS

I owe a lot to my wife Nirma and two daughters Maya and Amali, who exercise great tolerance to my division of time. My efforts to publicize Sri Lanka and its wildlife are made possible by the support of many people, especially my colleagues in Jetwing and others in the tourism industry. Past and present members of the Jetwing Eco Holidays team, including Ajanthan Shantiratnam, Aruni Hewage, L de S Gunasekera, Shehani Seneviratne, Chandrika Maelge, Ayanthi Samarajewa, Amila Salgado, Divya Martyn and Nirusha Ranjitkumar, have helped in numerous ways. My thanks to fellow Jetwing directors, who share my vision of the private sector supporting research and conservation. Hiran Cooray and Shiromal Cooray deserve thanks for supporting my unconventional multitasking techniques.

Chandra Jayawardana, Nadeera Weerasinghe, Anoma Algiyawadu, Hasantha Lokugamage 'Basha', Nilantha Kodituwakku, Dithya Angammana and Lal de Silva, present and past naturalists of Jetwing Hotels, assisted in the field. Chadraguptha Wickremesekera ('Wicky'), Supurna Hettiarachchi ('Hetti') and Chaminda Jayaweera and the other Jetwing Naturalist Chauffeur Guides deserve special mention for their assistance with my photography in the field. The field staffs of both the Forest Department and the Department of Wildlife Conservation have also assisted my photography, as well as many safari jeep drivers, especially those employed by Mola.

I could not have got this far in my efforts to popularize natural history if it were not for the support of my late parents Dalton and Lakshmi de Silva Wijeyeratne and my brother and sisters. My uncle Dodwell de Silva sparked off my interest in wildlife, and the interest in photography was encouraged by my aunt Vijita de Silva and my sister Manouri, who gave me my first cameras.

Various people helped with information generally or on specific topics. These include the late Dr Ravi Samarasinha, Rukshan Jayawardene, Andrew Kittle and Anjali Watson on Leopards, Dr Anna Nekaris and Lilia Bernede on primates, especially lorises, and on nocturnal observation techniques, Dr Shyamala Ratnayake (Sloth Bear), Rohan Pethiyagoda and Suyama Meegaskumbura (small mammals) and Rajith Dissainaike (squirrels). The family introductions to bats and primates are the 'long versions' of what was originally written for *Sri Lankan Wildlife* by Bradt Travel Guides. The cats introduction is based on a book on Leopards to be published by Jetwing Eco Holidays. I have used country introductory material previously used in books for New Holland. My thanks to Bradt, New Holland and Jetwing Eco Holidays for permission to use these sections.

I would also like to thank the other photographers who kindly contributed images to fill in gaps in my collection. They include Ruchira Somaweera, Suyama Meegaskumbura, Anna Nekaris, Neville Buck, Lilia Bernede-Beresford and Wipula Yapa. Rohan Pethiyagoda assisted in sourcing images.

My thanks to John Beaufoy and others in the New Holland team and to Krystyna Mayer for bringing the book into print and for being on the same wavelength as me for what this book should achieve.

INDEX